COPYRIGHT

Beyond the Algorithm: Building AI for the Good of Humanity

This book is a work of nonfiction. All research, examples, and references are presented for educational purposes. While every effort has been made to ensure accuracy, the author and publisher assume no responsibility for errors or omissions.

ABOUT THE AUTHOR

Alexandria Isaacs is a researcher whose work confronts one of the defining questions of our era: Will artificial intelligence elevate humanity or endanger it? Rooted in health technology and human-centered design, she has seen firsthand how powerful tools can heal, empower, and transform lives… and how easily they can be misused.

This duality, innovation's capacity for both good and harm, fuels her mission as an author. Alexandria writes not only to inform but also to influence: to shift public understanding, inspire readers to ask deeper questions, and encourage leaders to adopt principles that safeguard human dignity and freedom.

Alexandria writes to help readers understand the moral crossroads we now stand upon. Her work examines AI not just as a technological advancement but as a force that reflects human values, ethics, and intentions. Through clear analysis and balanced insight, she reveals how AI can be harnessed for compassion, fairness, and global wellbeing, or twisted toward manipulation, exploitation, and control.

Her books invite readers to become informed participants in shaping the future, reminding us that the true battle between good and evil will not be fought by machines, but by the humans who build and guide them

ABSTRACT

We stand at a crossroads. Artificial intelligence is no longer a distant dream or a figment of science fiction. It is woven into the fabric of our daily lives, shaping how we communicate, work, heal, and make decisions. Yet AI itself is neither savior nor destroyer. It is a mirror reflecting the values, biases, and choices of those who create and control it. This book is not about fearing technology; it is about understanding our responsibility in guiding it.

There is no doubt that artificial intelligence offers extraordinary potential, from diagnosing diseases earlier to extending human longevity, and from democratizing information to solving complex global challenges. But without ethics, compassion, and accountability, these same tools can erode privacy, distort truth, deepen inequality, and diminish human dignity.

The central premise of this book is simple but urgent: AI is not good or evil; humans are. The algorithms we build, the data we feed them, and the systems we deploy are all products of human intention. We are at a defining moment where our choices will echo for generations. Will we design AI that honors human dignity, or will we allow it to become a tool of manipulation and control? Will we prioritize fairness and transparency, or will we let bias and opacity flourish unchecked?

This book is written for everyone navigating this new landscape: professionals implementing AI in their organizations, policymakers crafting regulations, technologists writing code, and everyday people trying to understand how AI affects their lives. You don't need a computer science degree to engage with these questions. You need curiosity, concern, and a willingness to participate in shaping the future.

Throughout these chapters, we will explore real-world examples, practical frameworks, and actionable insights. We will examine how AI reflects humanity's capacity for both good and evil, how bias gets embedded into systems, and what responsible development truly looks like. We will confront uncomfortable truths about privacy, accountability, and the erosion of trust, but we will also discover pathways forward.

I hope that you finish this book not with fear, but with clarity and agency. We still have time to choose the world we want. But that choice requires awareness, intention, and collective action. Let's begin.

A Manifesto for the Future We Choose

We believe that artificial intelligence is the most powerful tool humanity has ever created.

A tool capable of healing, empowering, and elevating the human experience…

or diminishing it.

We believe AI is not good or evil on its own.

It reflects the values, intentions, and ethics of the humans who design and govern it.

Its impact will be shaped by our courage—or our complacency.

We believe humanity is at a crossroads.

One path leads to a future of compassion, accessibility, global healing, and shared progress.

The other leads to surveillance, manipulation, division, and the erosion of human dignity.

The choice is ours.

We believe the human spirit is irreplaceable.

No machine can replicate empathy, intuition, creativity, connection, or soul.

These gifts are ours alone—and they will matter more, not less, in the age of AI.

We believe in responsible innovation.

Technology should serve humanity, not exploit it.

AI must be transparent, ethical, and aligned with human well-being.

People—not power or profit—must remain at the center.

We believe everyone has a role to play.

Parents, teachers, caregivers, creators, citizens, and leaders.

Your voice matters.

Your choices matter.

Your awareness matters.

We believe the future is not predetermined.

We can shape it with intention, wisdom, compassion, and justice.

We can demand a future where AI is a force for good—because we are.

And we believe that humanity is strong enough to lead this transformation with heart.

The future will mirror us.

Let it mirror our very best.

TABLE OF CONTENTS

INTRODUCTION

Artificial intelligence has reached a turning point in human history, not because technology has achieved consciousness or autonomy, but because humanity now faces choices that will define the next century. The algorithms that recommend medical treatments, determine creditworthiness, screen job applications, and predict criminal behavior are not neutral arbiters; they are mirrors reflecting the values, biases, and priorities of the people who create and deploy them. This book examines a fundamental truth that often gets lost in discussions about AI's capabilities: technology itself carries no moral weight. The ethical implications of artificial intelligence rest entirely on human shoulders.

The question is not whether AI will shape society, but rather how humanity will shape AI. Every dataset selected, every optimization goal defined, and every deployment decision made represents a choice about whose interest's matter, whose dignity deserves protection, and what kind of world we want to inhabit. These choices are being made right now in corporate boardrooms, government agencies, research laboratories, and technology companies around the globe. The decisions being made today, often by small groups of developers and executives, will determine whether AI amplifies human flourishing or accelerates inequality, whether it protects privacy or enables unprecedented surveillance, and whether it strengthens democracy or facilitates manipulation.

The urgency of this moment cannot be overstated. Unlike previous technological revolutions that unfolded over generations, AI systems are being deployed at an extraordinary speed across virtually every sector of society. Healthcare algorithms are making treatment recommendations that affect patient outcomes. Facial recognition systems are being integrated into law enforcement despite documented accuracy gaps across demographic groups. Predictive analytics are shaping decisions about employment, housing, and education. Social media algorithms are curating the information that billions of people use to understand their world. Each deployment creates precedents, establishes norms, and generates data that trains the next generation of systems.

Yet despite AI's profound impact on daily life, most people feel disconnected from the decisions shaping these technologies. The perception persists that AI ethics is a technical domain reserved for computer scientists and philosophers, too complex for ordinary citizens to influence or understand. This perception is both false and dangerous. The ethical challenges posed by AI are fundamentally human questions about fairness, dignity, privacy, and power, questions that every person has the standing and capacity to address.

This book provides the frameworks, knowledge, and practical tools needed to engage meaningfully with AI ethics, whether as a professional implementing these systems, a leader making deployment decisions, a policymaker crafting regulation, or a citizen navigating an increasingly AI-mediated world. The chapters ahead explore how bias becomes embedded in algorithms, what responsible development practices look like, how privacy and security can be protected in an age of pervasive data collection, and who should govern AI systems and bear accountability when they cause harm, as well as how these technologies are transforming medicine and everyday life.

The exploration draws on documented cases, verified research, and established frameworks from organizations leading the development of AI ethics standards. The goal is not to inspire fear but to cultivate informed awareness, the kind of understanding that enables people to recognize when AI is being used without ethics, compassion, or responsibility, and to act in response.

Humanity still has time to choose the world it wants. The trajectory of AI is not predetermined. But that window of opportunity will not remain open indefinitely. The choices made in the coming years will either establish AI as a tool that serves human dignity and flourishing or allow it to become a mechanism for eroding privacy, fragmenting truth, increasing manipulation, deepening inequality, and threatening the fundamental rights that define free societies. The power to shape that outcome does not rest with algorithms. It rests with us.

CHAPTER 1.THE MIRROR: HOW AI REFLECTS HUMAN VALUES AND CHOICES

When a hospital algorithm systematically recommended less care for Black patients than for equally sick white patients, the problem wasn't the technology; it was us. Artificial intelligence had learned to replicate a healthcare system's historical pattern of spending less on Black patients faithfully mirroring a bias so deeply embedded that many humans hadn't consciously recognized it. This discovery wasn't an isolated incident; it was a revelation about the fundamental nature of artificial intelligence itself. AI systems don't create bias from nothing; they absorb it from us, learning to see the world through the lens of the data we feed them, the choices we make in designing them, and the values we embed, consciously or unconsciously, into their architecture.

The promise of artificial intelligence has always carried a seductive appeal: objective, data-driven decisions untainted by human emotion, prejudice, or irrationality. We imagine algorithms as neutral arbiters, mathematical entities that evaluate evidence without the messy complications of human psychology. This vision has driven the rapid adoption of AI across virtually every domain of modern life, from determining who gets hired and who receives medical treatment to predicting who might commit crimes and deciding which content billions of people see on social media platforms. Yet this faith in algorithmic objectivity rests on a fundamental misunderstanding of what AI is and how it comes to making decisions.

Every AI system begins with human choices. Developers choose which problems to solve and which to ignore. They select training data from a world already shaped by centuries of inequality, discrimination, and structural bias. They define success metrics that prioritize certain outcomes over others, efficiency over equity, profit over fairness, and measurability over human dignity. They make thousands of technical decisions about model architecture, feature selection, and optimization strategies, each one encoding assumptions about what matters and what doesn't. The algorithm that emerges from this process isn't a neutral tool; it's a crystallization of human values, a mirror reflecting the world we've created and the priorities we've chosen.

Understanding this reality is essential because the stakes have never been higher. AI systems now influence decisions that shape human lives in profound ways: whether someone receives a loan to buy a home, gets flagged as a security risk at an airport, qualifies for parole, or receives aggressive medical treatment versus palliative care. These aren't abstract technical questions; they're deeply human ones with consequences that ripple through families, communities, and entire societies. When AI systems perpetuate bias, they don't just make isolated errors; they systematize discrimination at scale.

automating inequality with unprecedented efficiency and cloaking it in the language of objectivity and scientific rigor.

This chapter explores how AI becomes a mirror of human values and choices by examining real-world cases where algorithms have faithfully reproduced our biases and sometimes amplified them beyond what their creators intended or imagined. From hiring algorithms that learned to discriminate against women by studying historical patterns in male-dominated industries to criminal justice tools that perpetuate racial disparities in sentencing to healthcare systems that allocate resources based on spending patterns that reflect existing inequities rather than actual medical need, these examples reveal a consistent pattern: AI doesn't solve our ethical problems; it forces us to confront them with new urgency.

The question isn't whether AI will reflect human values; it inevitably will. The question is which values it will reflect, whose interests it will serve, and whether we will make these choices consciously and deliberately or allow them to happen by default. Before we can build ethical AI systems, we must look honestly at ourselves, at the assumptions we carry, the blind spots we harbor, and the society we've constructed. Only by understanding how deeply human AI truly is can we begin the work of ensuring it reflects our highest aspirations rather than our worst tendencies. This is where the journey toward ethical AI must begin: with recognition that the mirror shows us exactly what we are.

The Illusion of Algorithmic Neutrality: Why AI Is Never Truly Objective

The belief that algorithms operate as neutral arbiters, free from the messy complications of human judgment, represents one of the most persistent and dangerous misconceptions about artificial intelligence. This illusion of objectivity stems from the mathematical nature of algorithms and their apparent distance from human emotion, yet it fundamentally misunderstands how AI systems come into being and function in the real world. Every algorithm is a human artifact, shaped by countless decisions made by its creators, trained on data generated by human activity, and deployed within social contexts already structured by power, inequality, and historical bias. **1 2**

The notion of algorithmic neutrality begins to crumble when examining how AI systems are constructed. Before a single line of code is written, developers make consequential choices about which problems deserve attention and resources. A company might invest millions in optimizing advertising while allocating minimal resources to detecting bias in hiring algorithms. These priority decisions reflect corporate incentives, market pressures, and societal values, not neutral mathematical principles. 1 2 The selection of what to build and what to ignore constitutes the first layer of human values embedded in AI systems.

Once a problem is selected, developers must define success. What does the algorithm aim to achieve? In criminal justice applications, should the system prioritize public safety, rehabilitation, fairness across demographic groups, or cost efficiency? Each objective encodes different values and produces different outcomes. A hiring algorithm optimized purely for employee retention might systematically exclude candidates who don't match historical patterns, even if those patterns reflect past discrimination. 1 2 The metrics chosen to measure algorithmic performance are never neutral; they represent judgments about what matters most. 1 2

The data used to train AI systems introduces another profound source of human influence. Machine learning algorithms learn patterns from historical data, which means they inevitably absorb the biases, inequalities, and prejudices embedded in that history. 2 When researchers at a major technology company developed an AI hiring tool trained on a decade of résumé data, the system learned to penalize résumés containing words associated with women, such as "women's chess club captain." The algorithm wasn't malfunctioning; it was accurately learning patterns from a dataset that reflected a male-dominated workforce. The company ultimately abandoned the tool, recognizing that historical data couldn't provide a neutral foundation for future decisions. 1 2

Even the technical architecture of AI systems reflects human choices. Developers select which features the algorithm should consider, how different variables should be weighted, and which mathematical approaches to employ. 1 2 These decisions often involve trade-offs between competing values. An algorithm designed to be highly accurate might be opaque and unexplainable, while a more transparent system might sacrifice some predictive power. Choosing between these options isn't a technical question with a single correct answer; it's an ethical judgment about which values take priority. 1 2

The deployment context further undermines claims of algorithmic neutrality. The same AI system can produce vastly different social outcomes depending on who controls it, how its outputs are interpreted, and what institutional structures surround its use. 1 2 A risk assessment algorithm in criminal justice doesn't simply generate scores; human judges decide whether and how to use those scores, often within systems already marked by racial disparities and unequal access to legal representation. 1 2

Perhaps most troubling, AI systems can create feedback loops that amplify existing biases. 1 2 When predictive policing algorithms direct officers to neighborhoods already subject to over-policing, they generate more arrests in those areas, which produces data suggesting those neighborhoods have higher crime rates, which reinforces the algorithm's original predictions. The system appears to validate itself while perpetuating and intensifying patterns of discriminatory enforcement. 1 2

Recognizing that algorithmic neutrality is an illusion doesn't mean AI is inherently harmful or should be abandoned. 1 2 3 Rather, it demands honesty about what AI systems actually are: powerful tools that reflect and amplify human values, priorities, and biases. 1 2 This recognition shifts the ethical question from whether AI can be neutral to whose values it should reflect and how those choices should be made. 1 2

Training Data as Cultural DNA: How Historical Biases Become Digital Realities

Training data functions as the foundational blueprint for artificial intelligence systems, encoding not merely statistical patterns but the accumulated historical biases, cultural assumptions, and value systems of the societies that produced it. 4 2 When AI systems learn from this data, they internalize centuries of human prejudice and systemic inequality, transforming historical injustices into algorithmic decisions that appear objective and neutral. This phenomenon represents one of the most consequential yet underrecognized challenges in AI ethics, because the biases embedded in training data translate directly into real-world harm affecting millions of people. 2

The problem extends far beyond simple technical oversight. Historical datasets reflect the priorities and blind spots of past eras, shaped by who had the resources to collect data, what was deemed worth measuring, and whose experiences were considered important enough to record. Medical research datasets, for decades, predominantly enrolled white male subjects, creating training data that encoded male physiology as the standard. When AI systems trained on this data are deployed in modern healthcare, they inherit this historical bias; women's symptoms are misclassified, dosing recommendations are mis calibrated, and diagnostic algorithms perform demonstrably worse for female patients. The bias originates not in the algorithm itself but in the data, it learned from, data that preserved historical inequities in medical research.

Criminal justice datasets used to train predictive policing and risk assessment algorithms contain records shaped by decades of discriminatory enforcement practices. 4 2 Over policing in certain neighborhoods created overrepresentation of those communities in arrest records. When AI systems learn from this data, they don't learn who commits crimes: they learn who gets arrested, a fundamentally different distribution. The algorithm then recommends increased surveillance in those same neighborhoods, creating a feedback loop that perpetuates and deepens historical injustices. 4 2

A critical danger lies in how AI systems transform historical correlations into causal predictions. Training data might show a correlation between zip code and loan default rates, a correlation that exists because of historical redlining, discriminatory lending practices, and systemic disinvestment in certain neighborhoods. An AI system trained on this data learns the correlation and uses the zip code as a predictive feature for loan decisions. The algorithm is not inventing discrimination; it is automating and legitimizing discrimination already embedded in the historical record. What makes this particularly insidious is the appearance of objectivity; the algorithm is simply following statistical patterns in the data, yet the outcome perpetuates historical injustice wrapped in the authority of algorithmic decision-making.

The scale and speed of AI amplify these historical biases in unprecedented ways. A biased hiring manager might discriminate against a few dozen candidates per year, but an AI system trained on biased data can make millions of biased decisions automatically, with no human intervention to catch or correct individual errors. The bias multiplies not through increased severity but through sheer volume and speed of deployment.

Perhaps most troubling is the feedback loop created when AI systems trained on biased historical data are deployed to make real-world decisions, and those decisions generate new data that feeds back into the next generation of AI systems. A hiring algorithm trained on biased historical data discriminates against certain groups. This discrimination becomes part of the historical record, the new data generated by the algorithm's decisions. The next generation of AI systems, trained on this expanded dataset, inherits both the original historical bias and the new bias introduced by the previous algorithm. The bias deepens with each iteration, documented in multiple domains from criminal justice to hiring to content moderation.

The Amazon Hiring Algorithm Case: When AI Learned to Discriminate Against Women

In **2014**, Amazon embarked on an ambitious project to revolutionize its hiring process through artificial intelligence, developing an automated system designed to screen résumés and identify top candidates with unprecedented efficiency. **5 7** The company's engineers trained the algorithm on a decade's worth of résumés submitted to Amazon, teaching it to recognize patterns associated with successful hires. **5 7 9** The system evaluated approximately **50,000** key terms across hundreds of résumé models, assigning candidates ratings from one to five stars based on how closely their applications matched historical patterns of success. **5 7** What emerged, however, was not an objective hiring tool but a digital manifestation of gender discrimination that would ultimately force Amazon to abandon the project entirely.

By 2015, Amazon's engineers discovered a troubling pattern: the algorithm systematically downgraded résumés from female applicants for technical positions. **5 6 7 8 9** The AI had learned to penalize résumés containing the word "women's," as in "women's chess club captain," or graduates of "women's colleges." **6 9** The system also favored résumés containing certain action verbs, such as "executed" and "captured," which data analysis revealed appeared more frequently in male applicants' materials. **6**

The discrimination wasn't a programming error or malicious intent; it was the algorithm faithfully learning from training data that reflected Amazon's predominantly male technical workforce, a pattern consistent across the broader technology industry. **5 6 9**

The mechanism of bias reveals how AI systems can transform historical inequality into automated discrimination. Amazon's technical workforce had been overwhelmingly male during the decade from which training data was drawn. **5 6 9** The algorithm identified this gender imbalance not as a problem to correct but as a pattern to replicate, essentially learning that being male correlated with hiring success. When presented with new résumés, the system applied this learned pattern systematically. Disadvantaging female candidates regardless of their actual qualifications or potential contributions. The AI was doing exactly what it was designed to do: identify patterns in historical data and use them to predict future outcomes, but those patterns encoded human bias rather than objective merit.

Amazon's engineers attempted to neutralize the bias by editing the algorithm to ignore explicitly gendered terms, but they could not guarantee the system wouldn't find other ways to discriminate. **5 7** The challenge illustrated a fundamental problem with machine learning: when training data reflects biased human decisions, the resulting algorithm inherits and automates that bias. Moreover, because the system used its own predictions to refine future results, it created a self-reinforcing feedback loop that entrenched gender discrimination rather than correcting it. **5** By **2017,** Amazon disbanded the team working on the project, with company representatives stating the tool was never used to make final hiring decisions, though internal reports indicated recruiters had consulted the AI's recommendations before the project's termination. **7 8**

The Amazon case demonstrates how AI systems can launder human bias through software, cloaking discrimination in the appearance of objective, data-driven decision-making. **6** Under **U.S.** employment law, algorithms that disproportionately disadvantage candidates based on protected characteristics like gender are illegal, regardless of whether the discrimination was intentional. **6** The case prompted increased scrutiny from the Equal Employment Opportunity Commission and renewed calls for algorithmic auditing and transparency in automated hiring systems. **6 8** It stands as a cautionary example of how AI, far from eliminating human prejudice, can systematize and scale discrimination with unprecedented efficiency when built on foundations of biased historical data, a mirror reflecting not the world as it should be, but as it has been, complete with all its inequities and exclusions. 5 6 8

Healthcare Algorithms and Racial Disparities: The Cost of Biased Predictions

In **2019**, researchers at the University of California, Berkeley, uncovered a troubling reality embedded in an algorithm used across the **U.S.** healthcare system to guide decisions affecting millions of patients. 15 17 The algorithm, designed to identify patients who would benefit from additional medical care and resources, systematically underestimated the healthcare needs of Black patients compared to equally sick white patients. 15 17 The bias was staggering in its impact: Black patients had to be significantly sicker than white patients, demonstrating **26.3%** more chronic illnesses, before the algorithm would recommend them for the same level of care. 13 This wasn't a minor technical glitch affecting a handful of cases. The biased algorithm influenced care decisions for approximately **200 million** people annually, and when researchers corrected the bias, the percentage of Black patients identified as needing additional care jumped from **17.7% to 46.5%.** 15

The mechanism of this bias reveals how seemingly neutral technical choices can encode and amplify racial disparities. The algorithm used healthcare costs as a proxy for medical needs, operating on the assumption that higher spending indicated sicker patients requiring more intensive intervention. 15 17 This assumption collapsed under the weight of a fundamental reality: the U.S. healthcare system has historically spent less on Black patients than on white patients with equivalent medical conditions. 10 14 15 Black patients face barriers to accessing care, receive fewer referrals to specialists, and encounter discrimination that results in less aggressive treatment. 10 14 15 When the algorithm learned from this data, it didn't learn who was sickest; it learned who had historically received the most expensive care, a pattern shaped by centuries of systemic racism in American medicine. 11 12 13 14 15 17

The consequences of such algorithmic bias extend far beyond statistical disparities. Patients denied access to care management programs miss interventions that could prevent complications, reduce hospitalizations, and improve quality of life. 12 14 15 A Black patient with multiple chronic conditions might not receive the coordinated care, medication management, or specialist referrals that a white patient with identical health status would receive, simply because an algorithm trained on biased data failed to recognize the severity of their condition. 12 14 15 The bias becomes self-perpetuating: inadequate care leads to worse health outcomes, which generates new data showing poorer health in Black populations, which algorithms then use to justify continued disparities in resource allocation. 13 14 17

Healthcare algorithms have demonstrated racial bias across multiple domains beyond care management. Risk prediction tools used to guide treatment decisions have systematically assigned different risk scores to patients based on race without scientific justification. 15 17 The American Heart Association's Get with the Guidelines, Heart Failure Risk Score, for example, assigned additional risk points to non-Black patients, effectively making it less likely that Black patients with equivalent illness severity would be referred to cardiologists. 13 Recent studies have documented that leading artificial intelligence platforms generate racially biased treatment recommendations for psychiatric patients, with the AI proposing different medication regimens based on race rather than clinical presentation. 16

The testing and diagnostic process itself reflects algorithmic bias. Research has found that medical testing rates for white patients are up to 4.5% higher than for Black patients with similar medical profiles. 10 This disparity in testing means that Black patients' health conditions are less thoroughly documented in medical records, creating training data that systematically underrepresents the true burden of illness in Black populations. **10 14 15** Algorithms trained on this incomplete data then perpetuate the cycle, recommending less testing and intervention for Black patients because historical records suggest they need less care, when the records simply reflect less care received. **10 14 15**

Addressing these disparities requires fundamental changes in how healthcare algorithms are designed, trained, and deployed. Researchers have demonstrated that replacing cost-based proxies with direct measures of disease burden can significantly reduce bias. **10 14** Diversifying training data, conducting rigorous audits for demographic disparities, and incorporating social determinants of health into algorithmic models represent essential steps toward equity. **11 14** Recent federal regulations now mandate greater transparency in predictive algorithms used in Medicare and Medicaid, providing mechanisms for scrutiny and accountability. **11** Yet the persistence of algorithmic bias in healthcare reveals a deeper truth: AI systems will continue to reflect and amplify the inequities embedded in the healthcare system itself until those underlying disparities are directly confronted and dismantled. **11 12 13 14 15 17**

Criminal Justice Risk Assessment Tools: Predicting the Future or Perpetuating the Past?

Across the United States, algorithms now influence some of the most consequential decisions in the criminal justice system: whether a defendant awaits trial in jail or at home, how long a sentence they receive, and when they become eligible for parole. **19** These risk assessment tools promise to bring objectivity and consistency to a system long plagued by human bias and arbitrary decision-making. Yet mounting evidence reveals a troubling paradox: algorithms designed to eliminate discrimination often encode and amplify the very inequities they were meant to solve, transforming historical injustice into mathematical formulas that appear objective and therefore become even more difficult to challenge.

The appeal of algorithmic risk assessment stems from genuine problems with human judgment. Research demonstrates that judges make more favorable parole decisions immediately after lunch than before it, illustrating how arbitrary factors influence critical determinations about human liberty. **18** AI systems operate without these irrational fluctuations, theoretically ensuring greater consistency. **18** They promise efficiency gains, directing costly detention resources toward cases with genuine public safety concerns while allowing lower-risk defendants to remain in their communities. **19** These benefits have driven widespread adoption; in many jurisdictions, algorithms have become standard practice for nearly every stage of the criminal justice process. **20**

The most damning evidence of algorithmic bias emerged from a **2016** ProPublica investigation of COMPAS (Correctional Offender Management Profiling for Alternative Sanctions), one of the most widely used risk assessment tools in the United States. The investigation revealed that the algorithm mislabeled Black defendants as high-risk nearly twice as often as white defendants, even when controlling past criminal behavior. **21** This finding shattered the assumption that removing race as an explicit variable ensures fairness. A comprehensive analysis by researchers examining over **750,000** bail decisions made in **New York** City between **2008** and **2013** found similar patterns; despite the algorithm not including race as a factor, it still exhibited significant biases due to the underlying data used to train it. **21**

The core problem extends beyond individual algorithmic design flaws. Risk assessment tools learn from historical crime data, using statistical methods to find patterns and connections. **20** These patterns, however, often represent existing inequities in the justice system rather than actual criminogenic factors. **20** The populations historically targeted by law enforcement, predominantly Black and Latino communities, subjected to decades of over-policing, face algorithmic scores that label them as likely to commit crimes, not because they are more criminal, but because they have been more heavily policed and prosecuted. [20] Data reflects social inequities even when variables such as gender, race, or sexual orientation are explicitly removed from the algorithm's inputs. [20] The bias is structural, embedded in the historical record itself.

This creates self-reinforcing feedback loops that deepen discrimination over time. When an algorithm trained on biased historical data makes biased predictions, those predictions influence new arrests and prosecutions, which then become new training data for the next iteration of the algorithm. [20] The bias compounds with each cycle, becoming increasingly pronounced and difficult to reverse. Some assessments perpetuate the very misconceptions and fears that drive mass incarceration, recommending detention for defendants who pose minimal risk while cloaking these decisions in the authority of data-driven objectivity. [20]

The stakes are extraordinarily high. Defendants held in pretrial detention are more likely to lose their jobs, face financial instability, and accept plea deals simply to gain their freedom, regardless of actual guilt. [21] An algorithmic error that results in unnecessary detention can fundamentally alter a person's life trajectory, causing job loss, family separation, and financial ruin, before any trial occurs. When a judge makes a biased decision, it can be appealed, scrutinized, and reversed. When an algorithm makes a biased decision, it is often treated as a technical output rather than a judgment subject to legal and ethical scrutiny, making discrimination appear objective and therefore more legitimate.

Design Choices as Moral Choices: How Programmers' Values Shape AI Behavior

Every decision a programmer makes while building an AI system, from selecting training data to defining successful metrics to choosing which features to prioritize, represents a moral choice with real-world consequences. These technical decisions are never neutral; they encode values, reflect priorities, and ultimately determine whose interests the system serves and whose it may harm. Understanding that design choices are moral choices is essential for anyone seeking to comprehend how AI systems come to exhibit the behaviors they do and why some algorithms perpetuate discrimination while others promote fairness.

The process of building an AI system requires thousands of individual decisions, each one shaping the final product in ways that extend far beyond technical performance. When developers choose which data to include in training sets, they make judgments about whose experiences matter and whose can be excluded. When they define what constitutes success for an algorithm, they prioritize certain outcomes over others, efficiency over equity, speed over accuracy, and profit over human welfare. When they decide how to handle edge cases or ambiguous situations, they determine which groups receive careful consideration and which are treated as statistical outliers unworthy of special attention.

Consider the seemingly straightforward task of building a content moderation algorithm for a social media platform. Developers must decide what constitutes harmful content, how to balance free expression against protection from abuse, and whether to prioritize removing all potentially harmful content or minimizing false positives that might censor legitimate speech. These decisions reflect deep value judgments about the relative importance of different harms, the role of platforms in shaping public discourse, and whose voices deserve amplification or suppression. A system designed to maximize user engagement will make fundamentally different choices than one designed to promote healthy conversation, and both will differ from one optimized to protect vulnerable users from harassment.

The values embedded in AI systems often remain invisible until they produce outcomes that reveal the priorities encoded within them. Research by Joy Buolamwini and Timnit Gebru demonstrated how facial recognition systems performed significantly worse on women and people with darker skin tones, with error rates reaching **34.7%** for dark-skinned women compared to less than **1%** for light-skinned men. This disparity didn't emerge from malicious intent but from design choices about training data collection, choices that treated lighter-skinned male faces as the default and failed to ensure diverse representation. The technical decision about which faces to include in training datasets was simultaneously a moral decision about whose accurate identification mattered.

The metrics developers choose to optimize reveal what they value most. An algorithm designed to minimize false positives will behave very differently from one designed to minimize false negatives, even when analyzing identical data. In medical diagnosis, optimizing to avoid false positives means the system errs on the side of caution, potentially missing some cases of disease to avoid unnecessary alarm. Optimizing to avoid false negatives means the system flags more potential cases, accepting more false alarms to ensure genuine cases aren't missed. Neither approach is inherently correct; the choice depends on value judgments about the relative costs of different types of errors, judgments that have profound implications for patient care and outcomes.

The composition of development teams shapes the values embedded in AI systems. Research consistently demonstrates that homogeneous teams produce systems with significant blind spots, failing to anticipate how their creations will affect people whose experiences differ from their own. When teams lack diversity across dimensions of race, gender, disability, socioeconomic background, and other characteristics, they are more likely to make design choices that work well for people like themselves while creating barriers or harms for others. The decision about who gets to build AI systems is itself a moral choice that determines whose values and perspectives shape the technology.

Recognizing design choices as moral choices demands greater accountability from developers and organizations. It requires moving beyond the notion that programmers are merely implementing technical specifications and acknowledging that they are making ethical decisions that affect millions of lives. This recognition opens space for more deliberate, transparent, and inclusive processes for making choices, processes that involve diverse stakeholders, consider multiple perspectives, and explicitly grapple with the value trade-offs inherent in AI development. The evidence presented throughout this chapter converges on a fundamental truth: artificial intelligence systems are not neutral mathematical entities operating beyond the realm of human influence, but rather sophisticated mirrors reflecting the values, biases, and choices of the people who create them and the societies that shape the data they learn from. The hospital algorithm that recommended less care for Black patients, Amazon's hiring tool that discriminated against women, and the criminal justice risk assessment systems that perpetuate racial disparities in sentencing all demonstrate the same underlying reality: AI amplifies and automates the patterns it finds in historical data, transforming centuries of human prejudice into algorithmic decision-making that affects millions of lives with unprecedented speed and scale.

This recognition fundamentally reframes the ethical challenge posed by artificial intelligence. The question is not whether AI can be made perfectly objective, because true objectivity is impossible when every stage of development involves human choices about what to measure, what to optimize, and whose experiences matter. The training data reflects historical inequities, the success metrics encode priorities about efficiency versus fairness, and the deployment contexts determine whether algorithmic outputs reinforce existing power structures or challenge them. Each technical decision, from which features to include in a model to how different types of errors should be weighted, represents a moral choice with real consequences for human dignity, opportunity, and justice.

The cases examined reveal patterns that extend across domains. Healthcare algorithms trained on cost data perpetuate racial disparities because they learn from a system that has historically spent less on Black patients. Criminal justice risk assessment tools predict future criminality based on past arrests, encoding decades of discriminatory policing into formulas that appear objective. Facial recognition systems perform poorly on women and people with darker skin because developers treated lighter-skinned male faces as the default. These failures share common roots in design choices that prioritized certain populations, perspectives, and outcomes while marginalizing others.

Understanding AI as a mirror of human values carries profound implications for how society approaches the development and deployment of these increasingly powerful systems. It demands recognition that the biases embedded in algorithms are not inevitable technical limitations but reflections of choices that can be made differently. It requires acknowledging that the people who build AI systems, their backgrounds, experiences, and blind spots, fundamentally shape what those systems become and whose interests they serve. It necessitates transparency about the value judgments encoded in technical decisions, creating space for democratic deliberation about whose values should guide AI development and how competing priorities should be balanced.

The stakes of getting this right have never been higher. AI systems now influence decisions about medical treatment, employment, criminal justice, financial opportunity, and access to essential services. When these systems inherit and amplify human bias, they don't merely perpetuate historical injustices; they systematize discrimination at scale, automating inequality with an efficiency and reach that no human decision-maker could achieve alone. Yet the mirror metaphor also offers hope: if AI reflects human values and choices, then different choices can produce different reflections. The algorithms of tomorrow need not replicate the biases of yesterday, but achieving this transformation requires confronting uncomfortable truths about the society reflected in today's data and making deliberate, sustained commitments to building systems that embody humanity's highest aspirations rather than its worst tendencies. The mirror shows exactly what we are and challenges us to become better.

CHAPTER 2: AI BIAS AND FAIRNESS: WHEN ALGORITHMS INHERIT OUR PREJUDICES

In **2018**, MIT researcher Joy Buolamwini discovered that commercial facial recognition systems could identify light-skinned men with near-perfect accuracy but failed to recognize dark-skinned women more than one-third of the time. This wasn't a random technical glitch; it was a systematic failure born from AI systems trained predominantly on datasets of white male faces, a digital manifestation of who had been considered the "default" human in technology development. In **2018**, MIT researcher Joy Buolamwini discovered that commercial facial recognition systems could identify light-skinned men with near-perfect accuracy but failed to recognize dark-skinned women more than one-third of the time. This wasn't a random technical glitch; it was a systematic failure born from AI systems trained predominantly on datasets of white male faces, a digital manifestation of who had been considered the "default" human in technology development.

The discovery sent shockwaves through the technology industry, but it shouldn't have surprised anyone paying attention to how artificial intelligence systems learn. These algorithms don't arrive in the world with prejudices; they acquire them from us, absorbing the biases embedded in the data we feed them, the choices we make in designing them, and the societal inequalities we've normalized to the point of invisibility. When we train AI systems on historical data that reflects centuries of discrimination, we shouldn't be shocked when those systems perpetuate and amplify the very patterns we claim to want to eliminate.

The challenge of algorithmic bias extends far beyond facial recognition technology. AI systems now influence decisions about who receives medical care, who gets hired for jobs, who qualifies for loans, who receives harsher criminal sentences, and who gains access to educational opportunities. In each of these domains, researchers have documented systematic disparities, algorithms that recommend less aggressive medical treatment for Black patients, hiring tools that downgrade applications from women, lending systems that charge higher interest rates to minority borrowers, and risk assessment tools that predict higher recidivism rates for defendants of color even when controlling for criminal history.

These aren't isolated incidents or unfortunate accidents. They represent a fundamental challenge at the intersection of technology and justice: how do we build fair AI systems when fairness itself is contested, when our training data reflects an unfair world, and when different mathematical definitions of fairness often contradict one another? The uncomfortable truth is that achieving algorithmic fairness requires us to confront uncomfortable questions about inequality in our society, questions we've often preferred to avoid or deny.

This chapter examines the complex mechanisms through which human prejudices become encoded in machine learning systems and explores what fairness might mean in algorithmic contexts. We'll investigate how bias enters AI through multiple pathways, not just biased training data, but also through the selection of proxy variables, the choice of optimization metrics, and the feedback loops that can turn small disparities into self-fulfilling prophecies. We'll explore why seemingly neutral data points like zip codes or word choices in job applications can serve as proxies for protected characteristics like race or gender, allowing discrimination to persist even when algorithms never explicitly consider those categories.

Understanding algorithmic bias matters because these systems increasingly shape the opportunities available to us and the treatment, we receive from institutions that hold power over our lives. When a facial recognition system fails to accurately identify people with darker skin, the consequences range from being inconvenient, being unable to unlock your phone, to dangerous, as when police use these systems to identify suspects and the technology's failures lead to wrongful arrests. When healthcare algorithms systematically underestimate the needs of certain populations, people suffer and die from treatable conditions. When hiring algorithms screen out qualified candidates based on gender or other characteristics, both individuals and organizations lose.

Yet this chapter also offers hope. By understanding how bias operates in AI systems, we can develop practical approaches to measure, mitigate, and monitor algorithmic fairness. We can build more diverse development teams, implement rigorous testing protocols, and create accountability structures that catch discriminatory patterns before they cause widespread harm. The path forward requires acknowledging that perfect fairness may be impossible, but that persistent, informed effort can create substantially more equitable systems than we have today, if we're willing to prioritize justice alongside efficiency and profit.

The Anatomy of Algorithmic Bias: How Prejudice Enters Machine Learning Systems

Understanding how bias infiltrates machine learning systems requires examining the entire lifecycle of AI development, from initial data collection through deployment and ongoing use. **23 24** Prejudice doesn't enter these systems through a single point of failure but rather through multiple pathways, each reflecting different aspects of human decision-making and societal inequality. **23 25 28**

The data collection stage represents the first and often most consequential entry point for bias. **23 24** Machine learning systems learn patterns from historical data, and when that data reflects a world shaped by discrimination, the algorithms faithfully reproduce those patterns. 23 25 28 A healthcare algorithm developed by Optum, used by hospitals across the United States, systematically recommended that Black patients receive less medical care than equally sick white patients. Researchers from the University of California, Berkeley discovered that the problem stemmed from using healthcare costs as a proxy for healthcare needs, since the American healthcare system has historically spent less on Black patients due to systemic barriers to access. The algorithm learned to associate being Black with needing less care, even when medical conditions were identical.

This example illustrates a critical mechanism of algorithmic bias: proxy discrimination. Even when algorithms never explicitly consider protected characteristics like race or gender, they can still achieve the same discriminatory outcomes through seemingly neutral variables that correlate with those characteristics. Zip codes encode information about racial composition due to residential segregation. Names signal gender and ethnicity. Educational institutions attended reflect socioeconomic background. When machine learning systems use these proxies to make predictions, they effectively discriminate based on protected characteristics while maintaining a veneer of objectivity.

The selection and labeling of training data introduce another layer of potential bias. 22 26 If datasets systematically underrepresent certain populations, algorithms will perform poorly for those groups. 26 27 Joy Buolamwini's research at MIT revealed that facial recognition systems achieved error rates exceeding thirty-four percent for dark-skinned women while maintaining near-perfect accuracy for light-skinned men. 24 26 This disparity was traced directly to training datasets composed predominantly of lighter-skinned faces, reflecting whose faces technology developers had considered important enough to include. 26

Human decisions about algorithm design and optimization also embed values and priorities into systems. 25 27 Every machine learning model requires developers to define what success looks like, which outcomes to maximize, which errors are acceptable, and how to balance competing objectives. These choices reflect implicit judgments about whose interests matter most. 22 A lending algorithm optimized purely-profit maximization might systematically deny loans to applicants from historically marginalized communities, even when those applicants would reliably repay, because the algorithm identifies patterns in historical data where discrimination created self-fulfilling prophecies about creditworthiness. 27

The feedback loops created by deploying biased systems represent perhaps the most insidious pathway for bias amplification. 23 When algorithms make decisions that affect the real world, those decisions generate new data that feeds back into future training. 23 Predictive policing algorithms trained on historical arrest data direct officers to patrol predominantly minority neighborhoods, leading to more arrests in those areas, which generates data suggesting those neighborhoods have higher crime rates, which reinforces the algorithm's recommendations. This cycle transforms initial bias into seemingly objective evidence; discrimination appears justified by data. 23

The validation and testing phase offers opportunities to detect bias but often fails to do so when evaluation datasets suffer from the same representational problems as training data. 22 If developers test facial recognition systems primarily on light-skinned faces, they won't discover the systems' failures with darker skin tones. If hiring algorithms are validated against historical hiring decisions made by biased humans, they'll appear successful even while perpetuating discrimination. 24

These multiple pathways through which prejudice enters machine learning systems share a common thread: they all originate in human choices and human-created conditions. [23 24 25 27] The data reflects the societies humans have built. [23] The proxies encode inequalities humans created and maintained. The optimization objectives prioritize the values humans selected. [22] The feedback loops amplify patterns humans initiated. [23] Recognizing this reality is essential because it means addressing algorithmic bias requires confronting the underlying human biases and structural inequalities that AI systems learn to replicate. [23 24 25 27]

The Fairness Impossibility Theorem: Why Different Definitions of Fairness Often Conflict

When ProPublica's investigative journalists analyzed the COMPAS recidivism prediction algorithm in 2016, they discovered that the system incorrectly flagged Black defendants as high-risk future offenders at nearly twice the rate it did for white defendants. [31 32] The algorithm's creators, however, insisted their system was fair, and mathematically, they were correct. [29] This apparent contradiction reveals one of the most challenging aspects of algorithmic fairness: different definitions of fairness are mathematically incompatible with one another, making it impossible for any single AI system to satisfy all fairness criteria simultaneously. [29 30 33]

This phenomenon, known among researchers as the fairness impossibility theorem, represents far more than an academic curiosity. It exposes a fundamental truth about justice in algorithmic systems: fairness is not a single, objective standard that can be calculated and optimized but rather a contested concept with multiple valid interpretations that often conflict. [29 30 33] The COMPAS algorithm satisfied one mathematical definition of fairness, calibration, meaning that among all defendants assigned a particular risk score, similar percentages of Black and white defendants reoffended. Yet it violated a different fairness criterion, equalized false positive rates, meaning that innocent Black and white defendants should be incorrectly labeled as high-risk at similar rates. **29**

The mathematical reality is stark: researchers have proven that these different fairness metrics cannot all be satisfied simultaneously except in trivial cases. An algorithm optimized for equal false positive rates across racial groups will necessarily produce different false negative rates. A system designed to ensure equal positive predictive value, meaning that a high-risk prediction is equally accurate across groups, cannot simultaneously guarantee equal false positive rates. These are not engineering challenges awaiting better solutions; they are mathematical certainties; trade-offs baked into the structure of prediction itself.

This impossibility extends beyond criminal justice to every domain where AI makes consequential decisions. Consider medical diagnosis algorithms: should fairness mean that the system achieves equal accuracy rates for all demographic groups, or that it identifies the same percentage of actual cases within each group, or that a positive diagnosis is equally reliable regardless of a patient's race or gender? These goals conflict mathematically. A hiring algorithm cannot simultaneously ensure that it selects the same percentage of qualified candidates from each demographic group, maintains identical false positive rates across groups, and produces recommendations that are equally predictive of job performance for all applicants.

The conflict deepens when considering individual fairness versus group fairness. **30 31** Individual fairness holds that similar people should receive similar treatment. This principle sounds uncontroversial until one confronts the challenge of defining similarity without reference to the very characteristics we're trying to protect against discrimination.31 Group fairness, by contrast, focuses on ensuring equitable outcomes across demographic categories; still, this approach requires collecting and using sensitive data about protected characteristics, which raises privacy concerns and potentially treats individuals differently based on group membership.31

Generative AI has intensified these tensions by forcing confrontation with a deeper question: should AI systems mirror the world as it exists, with all its historical inequalities, or should they generate outputs that reflect the more equitable world we aspire to create? **29.** If an AI system trained on historical data accurately reflects that most CEOs have been white men, is it being "fair" by reproducing that pattern or perpetuating injustice? If it deliberately generates more diverse representations, is it correcting bias or introducing distortion? **29**

These conflicts reveal that fairness in AI is fundamentally a matter of political and social negotiation, not merely a technical issue.29 Different stakeholders, including affected communities, developers, regulators, and businesses, hold competing values and priorities that cannot be reconciled through algorithms alone.29 The choice of which fairness metric to optimize represents a value judgment about whose interests matter most and what kind of equity we're trying to achieve. No mathematical formula can make that choice for us; it requires democratic deliberation, transparency about trade-offs, and ongoing adaptation as our understanding of justice evolves.

Proxy Discrimination: When Seemingly Neutral Data Points Encode Protected Characteristics

The most insidious form of algorithmic discrimination often hides in plain sight, disguised as neutral data points that seem to have nothing to do with race, gender, or other protected characteristics. When Amazon discovered in **2018** that its experimental hiring algorithm systematically downgraded applications from women, the system wasn't explicitly programmed to discriminate based on gender. Instead, it had learned to penalize résumés containing words like "women's" (as in "women's chess club captain") and graduates of all-women's colleges, seemingly neutral linguistic patterns that served as proxies for gender, allowing discrimination to persist even when the algorithm never directly considered whether an applicant was male or female.

Proxy discrimination occurs when AI systems use variables that correlate statistically with protected characteristics to make decisions, producing discriminatory outcomes without ever explicitly referencing race, gender, age, or other legally protected categories. **34 35 36 23 38** These proxies can be virtually anything: zip codes, educational institutions attended, first names, word choices in application essays, or even seemingly innocuous data like the time of day someone submits an online form. **35 36** The challenge lies in their ubiquity and unpredictability: because proxies can emerge from any variable that happens to correlate with protected traits in the training data, identifying and eliminating them becomes extraordinarily difficult. **35 36**

Zip codes represent perhaps the most recognized example of proxy discrimination. Due to decades of residential segregation, housing discrimination, and economic inequality, American zip codes encode substantial information about the racial and socioeconomic composition of neighborhoods. When lending algorithms use zip code as a factor in creditworthiness assessments, they effectively discriminate based on race without ever collecting data on applicants' racial identity. **35 37 23** The algorithm learns patterns from historical data showing that applicants from certain zip codes defaulted more frequently, patterns that reflect systemic barriers to economic opportunity rather than individual creditworthiness and perpetuate those disparities into future lending decisions.

Healthcare algorithms have demonstrated how proxy discrimination can have life-threatening consequences. Research published in science in 2019 revealed that a widely used healthcare algorithm serving millions of patients systematically underestimated the medical needs of Black patients. The algorithm used healthcare costs as a proxy for healthcare needs, operating on the seemingly reasonable assumption that sicker patients incur higher medical expenses. However, this proxy encoded centuries of racial disparities in healthcare access: Black patients historically received less medical care and generated lower costs not because they were healthier, but because systemic barriers prevented them from accessing the care they needed. **35.** The algorithm learned this pattern and recommended less aggressive care for Black patients even when their medical conditions were identical to those of white patients.

The technical challenge of addressing proxy discrimination stems from the fundamental nature of machine learning. **36 37 23** When researchers or regulators prohibit AI systems from using specific variables identified as proxies, algorithms can simply find alternative proxies that encode similar information. **34 38** Remove zip code from a lending algorithm, and it might rely more heavily on educational institutions attended or employment history, variables that also correlate with race and socioeconomic status. This phenomenon, sometimes called "proxy creep," means that eliminating individual proxies without addressing the underlying correlations in the data proves largely ineffective.

The legal framework for addressing proxy discrimination remains underdeveloped. Traditional anti-discrimination law prohibits explicit consideration of protected characteristics and, in some contexts, practices that produce unjustified disparate impacts. **34 36 38** However, proxy discrimination often falls into a gray area, particularly when the use of proxy variables can be justified by legitimate business purposes and when developers can claim they were unaware of the correlations their algorithms exploited. **34 36** The opacity of complex machine learning systems compounds this challenge; even developers may not fully understand which variables their algorithms weigh most heavily or how those variables relate to protected characteristics. **34 35 36**

Addressing proxy discrimination requires moving beyond simply removing obvious proxies to fundamentally rethinking how AI systems are designed, tested, and monitored. This means proactively analyzing training data for correlations between seemingly neutral variables and protected characteristics 38, implementing rigorous testing protocols that examine outcomes across demographic groups, and creating accountability structures that hold developers responsible for disparate impacts even when discrimination was unintentional. 34 35 38 The uncomfortable reality is that in a society marked by profound inequalities, virtually any data point can serve as a proxy for protected characteristics, meaning that building fair AI systems requires confronting the underlying inequalities that create those correlations in the first place.

The Feedback Loop Problem: How Biased AI Decisions Create Self-Fulfilling Prophecies

Perhaps the most troubling aspect of algorithmic bias lies not in the initial discriminatory decision, but in what happens next. When biased AI systems make predictions that influence real-world decisions, those decisions generate new data that feed back into future training, creating a vicious cycle where bias doesn't just persist, it intensifies. This phenomenon, known as the feedback loop problem, transforms algorithmic predictions into self-fulfilling prophecies that can entrench inequality for generations.

The mechanism operates with deceptive simplicity. An AI system trained on historical data makes a prediction about an individual or group. Decision-makers act on that prediction, altering the opportunities or treatment those individuals receive. The resulting outcomes, shaped by the prediction itself rather than underlying reality, become new data points that train future iterations of the algorithm. The system then interprets these outcomes as validation of its original prediction, reinforcing the pattern in subsequent decisions. What began as bias becomes encoded as apparent truth, backed by mounting evidence that the algorithm itself helped create.

Predictive policing systems demonstrate this cycle with stark clarity. Algorithms trained on historical arrest data identify neighborhoods with high crime rates and recommend increased police presence in those areas. Officers deployed to these neighborhoods naturally make more arrests, generating data that shows elevated crime rates in precisely the locations the algorithm predicted. This new data feeds back into the system, strengthening the algorithm's confidence that these neighborhoods require intensive policing. Meanwhile, areas with less historical police presence receive fewer patrols, resulting in fewer arrests regardless of actual criminal activity. The algorithm interprets this absence of arrests as evidence of lower crime rates, directing resources away from these neighborhoods. Over time, the initial bias, which may have reflected discriminatory policing practices rather than actual crime patterns, becomes increasingly difficult to distinguish from objective reality.

Research has documented how algorithmic predictions can alter behavior in ways that make those predictions come true, even when the initial assessment was unfounded. **39 41 42** When individuals learn that an algorithm has predicted negative outcomes for them, such as poor job performance, high credit risk, or likelihood of criminal behavior, this knowledge itself can influence their actions and opportunities. **39** A study on algorithmic assessments found that disclosing algorithmic predictions to evaluated individuals created anchoring effects, where people adjusted their beliefs and behaviors to align with the prediction, regardless of its accuracy. **39** Job candidates said that an algorithm predicted low performance subsequently performed worse, not because the algorithm identified genuine limitations, but because the prediction set low expectations that became self-fulfilling. **39**

The feedback loop problem proves particularly pernicious because it creates the illusion of algorithmic accuracy. As biased predictions shape outcomes that appear to validate those predictions, the system's performance metrics may improve over time, not because it's identifying truth, but because it's creating a reality that matches its biases. Decision-makers who rely on these apparently accurate predictions have no reason to question the system, and the cycle continues unchecked.

Breaking these feedback loops requires recognizing that algorithmic predictions are not passive observations but active interventions that shape the future they claim to forecast. This demands fundamental changes in how AI systems are designed, deployed, and monitored. Organizations must implement ongoing audits that examine not just whether predictions match outcomes, but whether the predictions themselves influenced those outcomes. **40** They must create mechanisms to inject external data that hasn't been shaped by previous algorithmic decisions, providing reality checks against self-reinforcing patterns. **40** Most importantly, they must acknowledge that in domains where AI systems influence the phenomena they predict, from criminal justice to employment to healthcare, accuracy metrics alone cannot determine whether a system is fair or whether it's simply perpetuating the biases it inherited.

Facial Recognition and the Accuracy Gap: Technical Failures with Real-World Consequences

Joy Buolamwini's **2018** research at MIT didn't just expose a technical flaw; it revealed a fundamental truth about how artificial intelligence systems can fail entire populations while appearing to work flawlessly for others. Her study of commercial facial recognition systems from major technology companies found that these algorithms achieved accuracy rates exceeding ninety-nine percent for light-skinned men but failed to correctly identify dark-skinned women more than thirty-four percent of the time. This wasn't a minor calibration issue or an acceptable margin of error. It was a systematic failure that rendered technology fundamentally unreliable for millions of people.

The technical explanation for this disparity traces directly to training data composition. Facial recognition algorithms learn to identify faces by analyzing thousands or millions of example images, extracting patterns about facial features, proportions, and variations. When these training datasets consist predominantly of lighter-skinned faces, particularly male faces, the algorithms become highly skilled at recognizing variations within that demographic while struggling to distinguish faces that fall outside their training experience. The National Institute of Standards and Technology's evaluation of facial recognition algorithms confirmed these disparities, finding that some systems demonstrated false positive rates for Asian and African American faces that were ten to one hundred times higher than for white faces. 45

The consequences of these technical failures extend far beyond inconvenience. In January 2020, Detroit police arrested Robert Williams, a Black man, based on facial recognition match that later proved incorrect. Williams spent thirty hours in detention, was fingerprinted and photographed, and experienced the trauma of being accused of a crime he didn't commit, all because an algorithm failed to accurately distinguish his face from a surveillance image. His case wasn't isolated. Subsequent wrongful arrests of Michael Oliver and Nijeer Parks, both Black men, followed similar patterns: poor-quality surveillance images fed into facial recognition systems that performed inadequately on darker-skinned faces, producing false matches that law enforcement treated as reliable evidence.

These failures occur despite impressive-sounding accuracy statistics that technology vendors promote. When companies claim ninety-nine percent accuracy, that figure typically represents performance under optimal conditions, high-quality images, controlled lighting, and frontal poses. Real-world deployment scenarios rarely match these conditions. The Center for Strategic and International Studies documented how an algorithm achieving a 0.1 percent error rate with high-quality photographs can see that error rate increases nearly one hundred-fold to

9.3 percent when processing images captured in uncontrolled environments. 43 Surveillance cameras produce grainy footage, subjects appear at angles, lighting varies dramatically, and faces may be partially obscured, all factors that degrade performance substantially.

The scale of facial recognition deployment amplifies these accuracy gaps into systemic problems. U.S. Customs and Border Protection has processed over three hundred million travelers using facial recognition technology. 44 Even if the system achieves 99.9 percent accuracy, an impressive figure in isolation, that 0.1 percent error rate translates to three hundred thousand potential misidentifications. When error rates correlate with demographic characteristics, these failures don't distribute randomly across the population but concentrate among already marginalized groups.

The industry has made genuine progress addressing demographic disparities. By 2024, NIST testing showed that leading algorithms achieved over ninety-nine percent accuracy across multiple demographic categories, with performance gaps narrowing substantially compared to earlier evaluations. 47 Yet this technical improvement doesn't resolve the fundamental ethical challenge: in high-stakes applications like law enforcement, border security, and access to essential services, even small disparities in accuracy translate into real harm for real people. A false match can result in wrongful arrest. 46 A false rejection can deny someone entry to their workplace or prevent them from accessing financial services. 46

Understanding facial recognition's accuracy gap requires looking beyond headline performance numbers to examine how these systems perform for specific populations in specific contexts. Technology may work remarkably well in controlled conditions while remaining dangerously unreliable in the messy reality of actual deployment, particularly for the communities most vulnerable to the consequences of algorithmic failure.

Measuring and Mitigating Bias: Practical Approaches to Building Fairer AI Systems

Recognizing bias in AI systems represents only half the challenge; the other half lies in developing practical methods to measure, reduce, and monitor algorithmic unfairness throughout the development lifecycle. While perfect fairness may remain mathematically impossible, substantial progress toward more equitable systems becomes achievable when organizations implement rigorous testing protocols, diverse evaluation metrics, and ongoing accountability structures.

Measuring bias requires moving beyond simple accuracy metrics to examine how AI systems perform across different demographic groups. **49** Researchers and practitioners have developed numerous fairness metrics, each capturing different aspects of equitable treatment. Statistical parity examines whether positive outcomes occur at similar rates across groups, for instance, whether loan approvals happen with equal frequency for applicants of different races. **52 53.** Equalized odds focus on whether the system achieves similar true positive and false positive rates across groups, ensuring that both opportunities and errors are distributed fairly. **49 52** Equal opportunities, a related metric, specifically examines whether qualified individuals from different groups receive positive outcomes at comparable rates. **49**

The disparate impact ratio, derived from employment discrimination law, provides a practical threshold: if an unprivileged group receives favorable outcomes at less than eighty percent the rate of a privileged group, the system may violate fairness standards. **49** This metric has proven particularly valuable in regulated domains like lending and hiring, where legal frameworks already establish expectations for equitable treatment.

Organizations building fairer AI systems increasingly employ specialized toolkits designed to detect and quantify bias. IBM's AI

Fairness 360 offers comprehensive capabilities for measuring bias across multiple metrics and implementing mitigation strategies. 48 Microsoft's Fair learn provides algorithms specifically designed to improve fairness in classification and regression models. 48 Google's What-If Tool enables interactive exploration of model predictions, allowing developers to visualize how outcomes vary across demographic subgroups and identify patterns of disparate treatment. 48

Data-centric approaches to bias mitigation begin before model training even starts. Ensuring training datasets represent all relevant populations prevents the underrepresentation that contributed to facial recognition failures. 48 50 Stratified sampling techniques deliberately balance dataset composition by including proportional representation from each demographic group. 48 When existing data proves insufficient, synthetic data generation methods like SMOTE (Synthetic Minority Over-sampling Technique) and ADASYN (Adaptive Synthetic Sampling) can augment underrepresented classes 48 50, though these approaches require careful validation to ensure synthetic examples reflect genuine population characteristics rather than introducing new distortions.

Algorithmic interventions during model training offer another layer of bias mitigation. Adversarial debiasing trains models alongside adversarial networks that attempt to predict sensitive attributes from the model's internal representations. 50 The primary model learns to make accurate predictions while simultaneously preventing the adversarial network from detecting protected characteristics, effectively removing correlations between predictions and attributes like race or gender. Fairness-aware training modifies loss functions to penalize disparate outcomes, encouraging models to optimize for both accuracy and equity simultaneously.

Post-processing techniques adjust model outputs after training is complete, offering flexibility when retraining proves impractical. Threshold adjustment sets different decision boundaries for each demographic group to satisfy fairness constraints, for example, adjusting the score required for loan approval to achieve equal opportunity across racial groups. **50** While effective at meeting specific fairness criteria, these approaches require careful consideration of whether differential treatment at the decision stage addresses underlying bias or introduces new forms of discrimination.

Human evaluation remains essential for detecting subtle biases that quantitative metrics miss. Adversarial testing deliberately probes models with inputs designed to expose prejudice, such as prompts containing stereotypes or edge cases involving marginalized groups. **50 51** Diverse review panels bring multiple perspectives to evaluating model outputs, catching context-specific unfairness that automated metrics overlook. **50** Continuous monitoring after deployment tracks model behavior in real-world conditions, incorporating user feedback and periodic audits to identify emerging bias patterns that weren't apparent during development. **51**

The most effective bias mitigation strategies combine multiple approaches across the AI development lifecycle, recognizing that no single intervention eliminates all forms of algorithmic unfairness. Organizations committed to building fairer systems implement bias detection and mitigation not as one-time fixes but as ongoing practices embedded in their development culture, accountability structures, and quality standards. **50 51** The discovery that facial recognition systems could identify light-skinned men with near-perfect accuracy while failing to recognize dark-skinned women more than one-third of the time, it wasn't just a technical problem; it was a mirror reflecting who had been considered important enough to include in technology development. This pattern extends across every domain where artificial intelligence makes consequential decisions about human lives. Healthcare algorithms that systematically underestimate the medical needs of Black patients, hiring systems that penalize applications from women, lending tools that charge higher interest rates to minority borrowers, and risk assessment instruments that predict higher recidivism for defendants of color all share a common origin: they learned from us, absorbing the biases embedded in historical data, design choices, and societal inequalities we've normalized to the point of invisibility.

The challenge of building fair AI systems proves far more complex than simply removing biased data or adjusting algorithmic parameters. The fairness impossibility theorem reveals that different mathematical definitions of fairness often contradict one another, making it impossible for any single system to satisfy all equity criteria simultaneously. An algorithm optimized for equal false positive rates across demographic groups will necessarily produce different false negative rates. A system designed to ensure equal accuracy for all populations cannot simultaneously guarantee that positive predictions are equally reliable across groups. These aren't engineering challenges awaiting better solutions; they are mathematical certainties that force uncomfortable choices about whose interests to prioritize and what kind of equity to pursue.

Proxy discrimination compounds these challenges by allowing bias to persist even when algorithms never explicitly consider protected characteristics like race or gender. Zip codes encode information about racial composition due to residential segregation. Names signal ethnicity and gender. Educational institutions attended reflect socioeconomic background. When machine learning systems use these seemingly neutral variables to make predictions, they effectively discriminate based on protected characteristics while maintaining a veneer of objectivity. The technical reality that removing individual proxies simply causes algorithms to find alternative proxies means that addressing discrimination requires confronting the underlying inequalities that create these correlations in the first place.

Perhaps most troubling, the feedback loop problem transforms initial bias into self-fulfilling prophecies that intensify over time. When biased AI systems make predictions that influence real-world decisions, those decisions generate new data that feeds back into future training. Predictive policing algorithms that direct officers to minority neighborhoods generate more arrests in those areas, creating data that appears to validate the original predictions. Healthcare systems that provide less care to certain populations generate cost data suggesting those populations need less care. What began as bias becomes encoded as apparent truth, backed by mounting evidence the algorithm itself helped create.

Yet understanding these mechanisms offers pathways toward more equitable systems. Organizations can implement rigorous testing protocols that examine performance across demographic groups and employ specialized toolkits like IBM's AI Fairness **360** and Microsoft's Fairlearn to detect and quantify bias. ensure training datasets represent all relevant populations and create accountability structures that catch discriminatory patterns before they cause widespread harm. The uncomfortable truth remains that achieving algorithmic fairness requires confronting uncomfortable questions about inequality in society, questions that extend far beyond technology into the realm of justice, power, and human dignity. Perfect fairness may prove impossible, but persistent, informed effort can create substantially more equitable systems than exist today if humanity chooses to prioritize justice alongside efficiency and profit. The algorithms reflect our choices, and those choices still belong to us.

CHAPTER 3: RESPONSIBLE AI DEVELOPMENT: BUILDING SYSTEMS WITH ETHICS AT THE CORE

Microsoft's **2016** chatbot Tay lasted less than **24 hours** on Twitter before the company pulled it offline, after the AI learned to spew racist and misogynistic content from its interactions with users who deliberately corrupted its training. The failure wasn't just a public relations disaster; it was a stark demonstration of what happens when developers deploy AI systems without adequate ethical safeguards, testing protocols, or consideration of how malicious actors might exploit vulnerabilities. Microsoft's **2016** chatbot Tay lasted less than **24 hours** on Twitter before the company pulled it offline, after the AI learned to spew racist and misogynistic content from its interactions with users who deliberately corrupted its training. The failure wasn't just a public relations disaster; it was a stark demonstration of what happens when developers deploy AI systems without adequate ethical safeguards, testing protocols, or consideration of how malicious actors might exploit vulnerabilities. The Tay incident became a watershed moment in technology circles, forcing uncomfortable conversations about responsibility, foresight, and the gap between what developers can build and what they should build.

The previous chapters explored how AI systems mirror human values and inherit our prejudices, revealing the profound ways that algorithmic bias perpetuates inequality. Understanding these problems represents a crucial first step, but awareness alone cannot prevent the next Tay, the next discriminatory hiring algorithm, or the next healthcare system that delivers unequal care. The question that emerges from this understanding is both urgent and practical: How do we build AI systems differently? How do we move from recognizing ethical failures after they occur to preventing them before they cause harm?

This chapter examines the frameworks, methodologies, and organizational practices that enable responsible AI development, the proactive integration of ethical considerations into every stage of the AI lifecycle. Rather than treating ethics as an afterthought or a compliance checkbox, responsible AI development demands that moral considerations shape initial design decisions, guide data collection and model training, inform testing protocols, and structure ongoing monitoring after deployment. This approach recognizes that technical excellence and ethical integrity are not competing priorities but interdependent requirements for AI systems that genuinely serve humanity.

The challenge extends beyond individual developers writing better code. Responsible AI development requires fundamental shifts in how organizations structure teams, allocate resources, define success metrics, and balance the pressure for rapid innovation against the necessity of careful ethical deliberation. It demands diverse perspectives around the table, not just computer scientists and engineers, but also ethicists, social scientists, domain experts, and representatives from communities that AI systems will affect. It requires transparency mechanisms that make algorithmic decision-making understandable to those impacted by it, accountability structures that assign clear responsibility when systems cause harm, and humility about the limitations of what we can predict or control.

Throughout this chapter, readers will encounter both successful implementations and cautionary tales that illuminate what responsible AI development looks like in practice. From algorithmic impact assessments that anticipate potential harm before deployment to participatory design processes that include affected communities in shaping AI systems, these approaches offer concrete pathways for embedding ethics into technology development. The frameworks presented here draw on established principles from organizations including the IEEE, European Union guidelines, and industry leaders who have pioneered ethics-by-design methodologies, adapted for practical application across different organizational contexts and technical domains.

The stakes of getting this right extend far beyond avoiding embarrassing headlines or regulatory penalties. As AI systems increasingly make decisions that affect employment, healthcare, criminal justice, financial opportunity, and countless other domains that shape human lives, the ethical choices embedded in their development will determine whether these technologies amplify human dignity or diminish it. Responsible AI development is not a destination to reach but an ongoing commitment, one that requires vigilance, continuous learning, and the courage to prioritize human welfare over efficiency, profit, or competitive advantage. The question facing every organization and individual involved in creating AI systems is not whether to embrace this responsibility, but whether they will do so before the cost of failure becomes unbearable.

Ethics by Design: Integrating Moral Considerations from Conception to Deployment

The concept of Ethics by Design represents a fundamental shift in how organizations approach AI development, moving ethical considerations from the periphery to the center of the creation process. Rather than treating ethics as a final checkpoint before deployment or a reactive response to public criticism, this methodology embeds moral deliberation into every phase of the AI lifecycle, from initial conception through ongoing monitoring. **54 55** The approach recognizes that ethical failures in AI systems rarely result from malicious intent but more often stem from inadequate foresight, insufficient stakeholder engagement, and the absence of systematic processes for identifying and addressing potential harms before they materialize.

The European Commission's guidelines on Ethics by Design articulate this proactive stance clearly: the methodology is intended to prevent ethical issues from arising in the first place by addressing them during the development stage, rather than attempting to fix problems after deployment. **55** This prevention-focused approach treats ethical principles not as abstract aspirations but as concrete system requirements that shape technical specifications, data governance practices, and organizational processes. **54 55 56** When developers at a European healthcare AI startup adopted this framework, early stakeholder workshops revealed concerns about training data bias and patient privacy that might otherwise have gone unaddressed until after deployment, potentially causing real harm to vulnerable populations. **54 55 56**

The practical implementation of Ethics by Design typically follows a structured, multi-phase process that integrates with existing development methodologies. **55 56** Organizations begin by establishing foundational ethical commitments aligned with recognized principles, including respect for human autonomy, fairness, transparency, accountability, privacy protection, and social well-being. **54 55 57** These high-level principles then inform scoping decisions as teams define the AI system's purpose, identify affected stakeholders, and analyze the socio-technical context in which the system will operate. **54** This preliminary work enables more rigorous impact assessments that examine foreseeable ethical, social, and legal risks before significant resources are committed to development.

During the design and implementation phases, ethical requirements translate into specific technical and organizational safeguards. 55 57 Privacy-by-design principles might mandate data minimization, encryption standards, and user consent mechanisms built into system architecture. Fairness requirements could necessitate diverse training datasets, bias detection algorithms, and regular audits of system outputs across demographic groups. Transparency commitments often result in explainability features that make algorithmic decision-making comprehensible to affected individuals, along with documentation practices that create clear audit trails. 55

The AI4 People Institute emphasizes that Ethics by Design represents an effective method for organizations to achieve not merely legal compliance but genuine trustworthiness, ensuring AI systems align with both the letter of regulations and the spirit of ethical responsibility. 54 This distinction matters because regulatory frameworks inevitably lag technological capabilities, and legal compliance alone cannot guarantee that AI systems serve human dignity and social good. Organizations committed to Ethics by Design accept responsibility for anticipating harms that regulations may not yet address and for upholding ethical standards that exceed minimum legal requirements.

Critical to this approach is the recognition that ethical integration cannot be a one-time event but requires continuous monitoring and iterative improvement throughout the system's operational life. 56 AI systems deployed in dynamic environments may encounter edge cases, shifting social contexts, or unintended consequences that only become apparent through real-world use. Responsible development therefore includes mechanisms for ongoing evaluation, stakeholder feedback, and system refinement that respond to emerging ethical concerns. This commitment to sustained vigilance acknowledges the limitations of what developers can predict or control at the outset, embracing humility about the complexity of deploying AI systems that affect human lives in consequential ways.

Building Diverse and Multidisciplinary AI Development Teams

The composition of AI development teams fundamentally shapes the ethical dimensions of the systems they create. When teams lack diversity in backgrounds, perspectives, and disciplinary expertise, they are more likely to produce AI systems that embed narrow worldviews, overlook critical ethical considerations, and fail to serve broad populations equitably. Research consistently demonstrates that homogeneous teams, whether in terms of demographics, educational backgrounds, or professional disciplines, are less effective at identifying biases, anticipating unintended consequences, and designing solutions that work for diverse user populations.

The case for diversity extends beyond demographic representation to encompass cognitive diversity and multidisciplinary collaboration. A team composed entirely of computer scientists, regardless of their demographic backgrounds, will approach AI development differently than one that includes ethicists, social scientists, domain experts, legal professionals, and community representatives. Each discipline brings distinct methodologies, questions, and concerns that enrich the development process. Ethicists can identify moral dilemmas that technical teams might frame purely as optimization problems. Sociologists can illuminate how AI systems might reinforce or disrupt existing power structures. Domain experts from healthcare, criminal justice, or education can provide crucial context about how algorithmic decisions affect real people in specific settings.

The benefits of diverse and multidisciplinary teams manifest throughout the AI development lifecycle. During the initial scoping phase, diverse perspectives help teams ask more comprehensive questions about who will be affected by the system, what values should guide its design, and what potential harm might emerge. A team developing a hiring algorithm that includes human resources professionals, labor advocates, and individuals who have experienced employment discrimination will surface concerns that a purely technical team might miss. During data collection and model training, team members with backgrounds in statistics, sociology, and the relevant application domain can collaboratively assess whether training data adequately represents affected populations and whether proxy variables might encode problematic biases.

Research on team composition in technology development reveals measurable advantages of diversity. Studies examining innovation outcomes have found that teams with greater demographic and disciplinary diversity generate more novel solutions and are better equipped to identify potential risks. Multidisciplinary teams are more likely to experiment with novel concepts and approaches to problems, an approach that can tie directly to pioneering advancements in AI. Organizations that have prioritized building multidisciplinary AI teams report improved ability to anticipate ethical challenges before deployment, more robust risk management processes, and stronger stakeholder trust. The financial services industry provides instructive examples: JPMorgan Chase's development of AI-powered fraud detection systems involved collaboration between data scientists, risk analysts, and compliance experts. resulting in systems that balanced accuracy with regulatory requirements and customer privacy protections.

Building genuinely diverse and multidisciplinary teams requires more than recruitment initiative; it demands organizational structures and cultures that value different forms of expertise. Performance evaluation systems must recognize and reward ethical deliberation, stakeholder engagement, and interdisciplinary collaboration, not just technical metrics like model accuracy or processing speed. Organizations need to create pathways for professionals from non-technical backgrounds to contribute meaningfully to AI development, providing necessary technical literacy without expecting them to become programmers.

Team structures should facilitate regular dialogue across disciplines, with sufficient time allocated for the kind of deliberative discussions that surface ethical concerns.

The challenge of assembling diverse teams is compounded by existing inequalities in technology fields. Women and people of color remain significantly underrepresented in computer science and AI roles, limiting the available talent pool. Addressing this reality requires both immediate strategies, actively recruiting from underrepresented groups, partnering with organizations that support diverse technologists, and creating inclusive workplace cultures, and long-term investments in education pipelines that broaden access to AI careers.

Ultimately, the composition of AI development teams is not merely a human resources consideration but an ethical imperative. The values, blind spots, and priorities of development teams become embedded in the systems they create. When those teams reflect narrow perspectives, the resulting AI systems will inevitably reflect those limitations, potentially causing harm to populations whose experiences and needs were never considered. Building diverse and multidisciplinary teams represents a practical, actionable step toward ensuring that AI systems serve humanity broadly rather than replicating existing patterns of exclusion and inequality.

Algorithmic Impact Assessments: Anticipating Harm Before It Occurs

Before the city government deployed an AI system to automate mortgage approvals, before a hospital implements an algorithm to prioritize patient care, before a company launches facial recognition technology in public spaces, there exists a critical window of opportunity to identify and prevent potential harms. Algorithmic Impact Assessments represent the systematic process of examining AI systems during this window, evaluating their potential consequences before deployment rather than scrambling to address damage after it occurs. **58 60**

An algorithmic impact assessment functions as a structured, preemptive review that analyzes how an automated decision system might affect individuals, communities, and broader societal values. **58 60 61** Unlike reactive approaches that respond to problems after they emerge, AIAs embed ethical deliberation into the development process itself, translating abstract principles like fairness and accountability into concrete requirements that shape system design. **58 60 66** The Canadian government's pioneering Algorithmic Impact Assessment tool exemplifies this approach, requiring public agencies to evaluate AI systems across multiple dimensions, including impact on rights and freedoms, procedural fairness, data quality, and privacy protection before deployment. **62**

The assessment process typically begins with comprehensive system mapping, documenting the AI system's purpose, functionality, intended users, and operational context. **60 61 62 64** Development teams must articulate clearly what decisions the system will make, what data it will use, and who will be affected by its output. **58 62** This foundational work enables more rigorous analysis of potential risks across technical, social, and ethical dimensions. **60 61 62 64** A mortgage approval algorithm, for instance, requires examination not just of its technical accuracy but of how historical lending data might encode decades of discriminatory practices like redlining, potentially causing the system to perpetuate rather than remedy existing inequalities. **58**

Effective AIAs demand multidisciplinary expertise and meaningful stakeholder engagement. **63 66** Technical teams can assess model performance and data quality, but identifying broader societal impacts requires input from ethicists, social scientists, domain experts, and, critically, representatives from communities that the system will affect. **63** When a U.S. city conducted an impact assessment for a proposed lending algorithm, community engagement revealed concerns about training data bias that technical teams had not fully appreciated, leading to fundamental redesign of the system before deployment. **58**

The assessment examines several critical dimensions that extend beyond immediate technical performance. Data quality evaluation scrutinizes training datasets for representativeness and latent biases that might produce discriminatory outcomes across demographic groups. **58 59 62** Risk analysis considers potential harms to rights, freedoms, equality, privacy, and well-being, not just for individuals but for communities and social systems. **58 60 62** Accountability structures assign clear responsibility for system outputs, particularly when failures occur. **58 60** Transparency mechanisms ensure that affected individuals can understand how decisions are made and access meaningful recourse when outcomes seem unjust. **58 62 65**

The European Union's AI Act has formalized impact assessment requirements for high-risk AI systems, reflecting growing recognition that such evaluations represent not merely best practices but essential governance mechanisms. **58 62** These regulatory frameworks acknowledge that the consequences of AI systems often extend far beyond their immediate technical functions, rippling through resource distribution, social equity, and democratic participation in ways that demand systematic anticipation and mitigation. **58 61**

Perhaps most importantly, effective AIAs recognize that impact assessment cannot be a one-time checkbox completed before deployment. AI systems operate in dynamic environments where edge cases emerge, social contexts shift, and unintended consequences become apparent only through real-world use.**58 60 61** Responsible development therefore includes mechanisms for ongoing monitoring and periodic reassessment, creating feedback loops that enable continuous improvement and rapid response to emerging harm. **60 61**

The challenge facing organizations developing AI systems is not whether to conduct impact assessments but whether they will do so with sufficient rigor, genuine stakeholder engagement, and commitment to acting on findings even when mitigation requires difficult trade-offs. **63 66** The window of opportunity to prevent harm closes quickly once systems are deployed at scale, making the discipline of anticipating consequences before they occur not just ethically sound but practically essential. **58 60**

Transparency and Explainability: Making AI Decision-Making Understandable

When a bank's AI system denies a mortgage application, when a hospital algorithm recommends against a particular treatment, when a hiring platform filters out a candidate, the individuals affected by these decisions face a fundamental question: Why? The inability to answer this question represents one of the most pressing challenges in responsible AI development. Transparency and explainability have evolved from technical considerations to ethical and legal imperatives as AI systems increasingly make consequential decisions that shape human lives. **1 2**

Transparency addresses the visibility of AI system operations: what data the system uses, how it processes information, and what factors influence its outputs. Explainability focuses on making the reasoning behind specific decisions comprehensible to both technical and non-technical stakeholders. **1** 2 Together, these principles enable accountability, facilitate bias detection, and respect individual autonomy by allowing people to understand and potentially contest decisions that affect them.

The challenge emerges from the fundamental architecture of modern AI systems. Deep learning models, particularly the neural networks that power contemporary applications, function as "black boxes" where the relationship between inputs and outputs remains opaque even to their developers. 2 While traditional machine learning models like decision trees provide transparent decision pathways, contemporary AI systems prioritize predictive accuracy at the expense of interpretability. 1 2 A model might correctly predict loan default risk with impressive accuracy yet offer no comprehensible explanation for why it flagged a particular applicant as high-risk.

This opacity creates significant accountability gaps. When AI systems make errors or produce discriminatory outcomes, organizations deploying these systems cannot readily explain to users, regulators, or affected individuals why specific decisions were made. 2 The European Union's AI Act has begun addressing this challenge by establishing transparency obligations for high-risk AI applications, requiring organizations to maintain and provide documentation of AI system operations. 1 2 Similarly, data protection regulations increasingly require organizations to explain automated decision-making to affected individuals, reflecting recognition that transparency serves both accountability and individual rights. 1 2

Several technical approaches have emerged to address the explainability challenge. Methods such as LIME (Local Interpretable Model-agnostic Explanations) and SHAP (SHapley Additive explanations) provide post-hoc explanations by analyzing how model predictions change when inputs are modified. In transformer-based models, attention mechanisms can reveal which input components the model prioritized when generating outputs. Feature importance analysis identifies which variables most strongly influence model decisions. Yet even when these technical methods function correctly, a gap often persists between technical explanations and human understanding. 1 2 A feature importance ranking may be technically accurate yet fail to convey meaningful insight about why a specific decision occurred.

Achieving meaningful transparency requires integrating explainability considerations throughout the AI development lifecycle, from problem formulation and data selection through model training, validation, and deployment. **1 2** Organizations must establish governance structures that prioritize transparency as a core value rather than a compliance checkbox. This integration demands cross-functional collaboration between data scientists, ethicists, legal professionals, and domain specialists who can translate technical complexity into comprehensible explanations for varied audiences. **1**

Comprehensive documentation of AI systems, including design decisions, training data characteristics, model limitations, and intended use cases, creates the foundation for transparency and auditability. **2** model cards and datasheets for datasets enable internal review, external auditing, and regulatory oversight. Such documentation also supports organizational learning by preserving institutional knowledge about the specific technical choices that were made.

The tension between transparency demands and competitive pressures remains unresolved. Organizations developing proprietary AI systems often resist full disclosure of model architectures or training data, citing secret trade protections. 1 Balancing these competing interests, ensuring sufficient transparency for accountability while protecting legitimate commercial interests, represents an ongoing challenge in responsible AI governance. 1 Yet the alternative, deploying opaque systems that make consequential decisions without explanation, undermines the accountability structures necessary for AI systems to serve human dignity rather than diminish it.

Participatory Design and Stakeholder Engagement: Including Affected Communities

The most sophisticated algorithmic impact assessment, the most diverse development team, and the most rigorous technical safeguards can still fail to prevent harm if they exclude the voices of those who will live with the consequences. Participatory design and stakeholder engagement represent a fundamental shift in how AI systems are created, moving from a model where technical experts make decisions on behalf of affected communities to one where those communities actively shape the systems that will impact their lives. **4 2 67**

Participatory design in AI development refers to the systematic inclusion of diverse stakeholders, particularly those from marginalized or affected communities, throughout the entire lifecycle of system creation, from initial conception through deployment and ongoing evaluation. **2 67** This approach recognizes that people who will be subject to algorithmic decisions possess crucial knowledge about their own contexts, needs, and vulnerabilities that technical teams, regardless of their expertise, cannot fully anticipate from the outside. **2** When a hospital system develops an AI triage tool without consulting emergency department patients, clinicians, and community health advocates, it risks creating a system that optimizes for metrics that matter to administrators while overlooking factors that determine whether patients receive appropriate, dignified care.

The ethical foundation for participatory approaches rests on principles of autonomy, justice, and respect for human dignity. Communities affected by AI systems have a moral right to influence decisions that shape their opportunities, freedoms, and well-being. **67** This principle appears in professional codes, including the Association for Computing Machinery's Code of Ethics, which emphasizes that computing professionals must ensure stakeholders have opportunities to participate in decisions that affect them. **67** Beyond rights-based arguments, participatory design offers practical benefits that improve system quality and effectiveness. Engagement with affected communities helps identify sources of bias that homogeneous development teams might overlook **67, 68**, provides context-specific insights that increase system appropriateness, and builds trust that facilitates adoption and reduces resistance 2 67.

Effective stakeholder engagement requires more than token consultation or superficial feedback sessions. Meaningful participation demands that affected communities have genuine influence over substantive decisions, not merely cosmetic choices about user interface design but fundamental questions about whether a system should exist, what problems it should address, what trade-offs are acceptable, and how accountability will be ensured. 67 The city of Amsterdam's approach to AI-powered urban safety monitoring illustrates this principle in practice. Rather than deploying surveillance technology and seeking community input afterward, local authorities partnered with residents and advocacy groups to co-design the system from inception, allowing community members to define acceptable uses, establish privacy safeguards, and shape redress mechanisms. 67 This collaborative process directly influenced system requirements and governance structures.

The methods for participatory design vary depending on context but typically include stakeholder mapping to identify all affected parties 2 67, workshops and focus groups that elicit community priorities and concerns, co-design sessions where stakeholders contribute directly to system specifications, and continuous feedback mechanisms that enable ongoing input throughout development and deployment 2 67. These processes must address power imbalances that can undermine genuine inclusivity, ensuring that marginalized voices are not overshadowed by more dominant stakeholders 67 and that technical complexity does not create barriers to meaningful participation 2 67.

The challenges of participatory design are significant. Meaningful engagement requires time, resources, and organizational commitment that may conflict with pressures for rapid development and deployment. 2 non-technical stakeholders may need accessible information and facilitation to engage effectively with complex technical issues. **2 67** Organizations must guard against tokenism, superficial inclusion that creates the appearance of participation without transferring real decision-making power. **2 67**

Yet the alternative, developing AI systems that profoundly affect communities without their input, perpetuates patterns of exclusion and increases the likelihood of systems that cause harm, erode trust, and fail to serve the populations they purport to help. Participatory design acknowledges that those who will live with AI systems possess essential knowledge for creating technology that genuinely serves human dignity rather than diminishing it. **2 67 68**

Continuous Monitoring and Accountability: Ensuring Ethical Performance Over Time

Deploying an AI system represents not an endpoint but a beginning, the start of an ongoing relationship between technology and the communities it affects. The most rigorous pre-deployment testing, the most comprehensive impact assessments, and the most carefully designed safeguards cannot anticipate every scenario an AI system will encounter in the complex, dynamic environments where it operates. Continuous monitoring and accountability mechanisms transform responsible AI development from a one-time achievement into a sustained commitment, ensuring that systems maintain ethical performance throughout their operational lives. **70, 71, 72.**

The necessity of continuous monitoring emerges from fundamental characteristics of AI systems and the environments in which they function. Machine learning models trained in historical data may encounter distribution shifts as real-world conditions change, potentially degrading performance or producing unexpected outcomes **69**. Biases that were undetectable during testing may become apparent only when systems operate at scale across diverse populations **69, 70, 74.**

New use cases may emerge that developers never anticipated, raising ethical questions that initial assessments did not address 71, 72. Without systematic monitoring, these issues can persist undetected, causing cumulative harm that erodes trust and undermines the legitimacy of AI applications.

Financial institutions deploying AI-powered surveillance tools to detect fraud and market manipulation have pioneered approaches to continuous monitoring that offer instructive models for other sectors 69. These organizations regularly evaluate model performance in real-world applications, actively seeking to identify emerging biases through ongoing analysis of system outputs across different customer segments 69. Human analysts review flagged cases to assess whether the AI system's decisions align with regulatory requirements and ethical standards, creating feedback loops that enable continuous model refinement 69. This iterative process acknowledges that maintaining system integrity requires vigilance rather than if initial deployment testing provides permanent assurance.

Effective monitoring encompasses multiple dimensions that extend beyond technical performance metrics. Organizations must track not only accuracy and efficiency but also fairness outcomes across demographic groups, examining whether the system produces disparate impacts that violate equity principles 70, 74. Privacy monitoring ensures that data handling practices remain compliant with regulations and respect individual rights as systems process information at scale 69, 71, 72. Security assessments identify vulnerabilities that malicious actors might exploit, particularly as adversarial techniques evolve 69, 71, 72. Transparency mechanisms enable external scrutiny, allowing regulators, researchers, and affected communities to evaluate whether systems operate as intended. 70

Accountability structures provide the organizational foundation that transforms monitoring data into meaningful action. Clear assignment of responsibility ensures that specific individuals or teams bear obligation for system oversight, performance evaluation, and response to identified problems **73.** Ethics committees or review boards can provide independent assessment of monitoring findings, particularly when trade-offs between competing values require deliberation that extends beyond technical expertise **69.** Documentation practices create audit trails that enable retrospective analysis when failures occur, supporting both organizational learning and external accountability to regulators or affected parties. **70, 73**

The European Union's AI Act and similar regulatory frameworks increasingly mandate continuous monitoring and accountability mechanisms for high-risk AI applications, reflecting recognition that ongoing oversight represents not merely best practice but essential governance **70.** These requirements acknowledge that the dynamic nature of AI systems and their deployment contexts create risks that emerge over time rather than manifesting immediately at deployment. **71 72.** Organizations that treat monitoring as compliance theater, conduct perfunctory reviews that lack genuine commitment to identifying and addressing problems, miss the fundamental purpose of these practices.

Human oversight remains indispensable even as monitoring processes incorporate automated tools **69, 70, 71.** Algorithms can flag statistical anomalies or performance degradation, but human judgment is necessary to interpret findings within broader ethical and social contexts, to weigh competing values when trade-offs arise, and to make decisions about system modification or decommissioning when monitoring reveals unacceptable harms. This human-in-the-loop approach recognizes that accountability ultimately rests with people, not algorithms, and that the most sophisticated monitoring systems serve to inform rather than replace human responsibility for ensuring AI systems serve human dignity. The journey from recognizing AI's ethical challenges to building systems that embody ethical principles require more than good intentions; they demand systematic approaches.

organizational commitment, and sustained vigilance. Responsible AI development is not a destination that organizations reach through a single initiative or compliance checklist, but rather an ongoing practice that must be woven into the fabric of how teams conceive, design, test, deploy, and maintain AI systems. The frameworks and methodologies explored throughout this chapter, from ethics by design to continuous monitoring, represent practical pathways for translating abstract ethical principles into concrete technical and organizational practices that shape how AI systems function in the world.

The evidence from both successful implementations and cautionary failures reveals several essential truths about responsible AI development. First, ethical considerations cannot be bolted onto AI systems as afterthoughts without fundamentally compromising their integrity. When Microsoft deployed Tay without adequate safeguards, when hiring algorithms discriminated because developers failed to examine training data for bias, and when healthcare systems perpetuated racial disparities through algorithmic recommendations, these failures stemmed not from malicious intent but from treating ethics as peripheral rather than central to the development process. Organizations that successfully build responsible AI systems recognize that ethical deliberation must inform initial scoping decisions, shape data governance practices, guide model selection and training, and structure ongoing evaluation throughout the system's operational life.

Second, the composition of development teams profoundly influences the ethical dimensions of the systems they create. Homogeneous teams, regardless of their technical sophistication, lack the diverse perspectives necessary to identify biases, anticipate unintended consequences, and design solutions that serve broad populations equitably. Multidisciplinary collaboration brings ethicists, social scientists, domain experts, and community representatives into dialogue with technical teams, enriching the development process with questions and insights that purely technical approaches would miss. The inclusion of affected communities through participatory design processes ensures that those who will live with AI systems have genuine influence over the decisions that shape their opportunities, freedoms, and well-being.

Third, transparency and accountability mechanisms transform abstract responsibility into concrete organizational practices. Algorithmic Impact Assessments create structured processes for anticipating potential harm before deployment rather than scrambling to address damage after it occurs. Documentation practices, explainability requirements, and continuous monitoring establish the foundations for both internal oversight and external accountability to regulators and affected communities. These mechanisms acknowledge that developers and deploying organizations cannot predict every scenario an AI system will encounter, making ongoing vigilance essential rather than optional.

The tension between rapid innovation and careful ethical deliberation remains unresolved in many organizations. Competitive pressures, resource constraints, and the allure of being first to market create powerful incentives to shortcut the deliberative processes that responsible AI development requires. Yet the costs of ethical failures, in human harm, eroded trust, regulatory penalties, and damaged reputations, increasingly demonstrate that speed without responsibility represents a false economy. Organizations that prioritize responsible development practices are discovering that ethical integrity and technical excellence are not competing priorities but interdependent requirements for AI systems that genuinely serve humanity.

The frameworks presented in this chapter offer concrete starting points, but their effectiveness depends entirely on the commitment of organizations and individuals to implement them with genuine rigor rather than treating them as compliance theater. As AI systems continue to proliferate across healthcare, criminal justice, employment, financial services, and countless other domains that shape human lives, the ethical choices embedded in their development will determine whether these technologies amplify human dignity or diminish it. The question facing every organization and individual involved in creating AI systems is not whether to embrace this responsibility, but whether they will do so with the serious and sustained effort that the stakes demand, before the window of opportunity to prevent harm closes and the costs of failure become unbearable.

CHAPTER 4: AI PRIVACY AND SECURITY: PROTECTING HUMAN DIGNITY IN THE DIGITAL AGE

Target, the retail giant, knew a teenager was pregnant before her father did; their AI system had analyzed her shopping patterns and detected the telltale combination of unscented lotion, magnesium supplements, and cotton balls that predicted pregnancy with remarkable accuracy. When the father angrily confronted the store about sending maternity coupons to his high school daughter, only to apologize days later after learning she was indeed pregnant, the incident revealed how AI can extract our most intimate secrets from seemingly innocuous data trails we leave behind. This incident, which occurred in **2012**, wasn't an isolated case of corporate overreach; it was an early warning of how artificial intelligence would fundamentally transform the relationship between individuals and the organizations that collect data about them. The teenager's shopping habits, seemingly mundane purchases made with a loyalty card, became data points that an algorithm analyzed to reveal one of life's most private moments. What makes this revelation particularly unsettling is not just that Target knew, but that the knowledge came from patterns invisible to human observation, extracted by machine learning systems designed to predict behavior with eerie precision.

Privacy has always been essential to human dignity, providing the space for authentic self-expression, intimate relationships, and the freedom to make choices without constant scrutiny or judgment. Yet AI-powered technologies are dismantling this fundamental right at an unprecedented scale and speed. Every digital interaction, from the websites we visit to the pace at which we type, from our location data to our heart rate patterns captured by wearables, generates information that sophisticated algorithms can analyze to infer our health conditions, political beliefs, financial stability, sexual orientation, and psychological vulnerabilities. These inferences happen invisibly, without our explicit consent or even our awareness, creating detailed profiles that shape the opportunities we're offered, the prices we pay, and the information we see.

The erosion of privacy in the AI age represents more than lost confidentiality. When algorithms can predict our behavior better than we can predict it ourselves, when surveillance systems track our movements through public spaces, and when our most intimate health data becomes fodder for predictive analytics, we lose something essential to human autonomy: the capacity to define ourselves rather than being defined by our data shadows. The question is no longer whether organizations have information about us, but whether we retain any meaningful control over how that information is collected, analyzed, and used to make decisions that affect our lives.

This chapter explores how AI amplifies privacy threats in ways that previous technologies never could, examining the mechanisms through which machine learning systems extract sensitive information from seemingly innocuous data, the surveillance infrastructure that increasingly monitors our daily lives, and the vulnerabilities that emerge when AI analyzes health information. The stakes extend beyond individual privacy violations to encompass threats to democratic participation, freedom of expression, and the basic human right to live without constant monitoring and behavioral manipulation.

Yet this chapter also offers hope and practical guidance. Privacy-preserving technologies are emerging that allow us to benefit from AI's analytical power without sacrificing personal information. Regulatory frameworks like the European Union's General Data Protection Regulation establish new rights and protections, demonstrating that privacy erosion is not inevitable. Organizations and individuals can make choices that prioritize human dignity over data extraction, implementing technical safeguards and demanding transparency about how AI systems use personal information.

Understanding how AI threatens privacy is the first step toward protecting it. The algorithms that analyze our shopping patterns, track our faces in crowds, and predict our health conditions operate according to design choices made by humans, choices that can be challenged, regulated, and changed. As we examine the privacy landscape transformed by artificial intelligence, we must remember that technology reflects human values and priorities. The question is not whether AI will respect privacy, but whether we will demand and build systems that do. The teenager whose pregnancy Target predicted deserved privacy not because perfect confidentiality is technically achievable, but because human dignity requires spaces free from algorithmic surveillance and inference. That principle remains our guide as we navigate the complex terrain where artificial intelligence meets personal privacy.

The Privacy Paradox: How AI Extracts Intimate Knowledge from Everyday Data

The paradox of privacy in the age of artificial intelligence lies in a troubling contradiction: people express genuine concern about protecting their personal information yet routinely surrender intimate details of their lives through everyday digital interactions. 1,4 This behavioral inconsistency, documented extensively by privacy researchers, reveals how AI systems have become extraordinarily adept at extracting sensitive knowledge from data that appears mundane on its surface. The teenager whose pregnancy Target predicted through purchase patterns represents just one example of how machine learning algorithms can infer our most private circumstances from the digital breadcrumbs we scatter without conscious thought.

Every interaction with digital technology generates data that AI systems can analyze to reveal far more than users intend to disclose. 1 2 Location data from smartphones reconstructs not just where people go, but patterns that expose religious affiliations, political activities, romantic relationships, and health conditions based on visits to places of worship, campaign offices, specific addresses, or medical facilities. Fitness trackers and health monitors collect biometric information, heart rate variability, sleep patterns, and activity levels that algorithms can analyze to detect early signs of depression, pregnancy, or chronic illness before individuals themselves recognize symptoms. Social media platforms harvest not just explicit posts and like but metadata about when users are active, how quickly they respond to certain content, and the network structure of their connections, all of which feed algorithms that build detailed psychological profiles. 1 2

The power of AI to extract intimate knowledge stems from its capacity for inferential analytics, the ability to predict sensitive characteristics without direct disclosure. Research has demonstrated that machine learning models can accurately predict sexual orientation from facial photographs, determine political ideology from Facebook likes, and assess personality traits from smartphone usage patterns. These inferences happen invisibly, without users' awareness or explicit consent, creating what privacy scholar Helen Nissenbaum calls violations of "contextual integrity," the appropriate flow of information within specific social contexts.

Smart home devices present particularly acute privacy vulnerabilities. 1 2 Voice assistants continuously listen for activation commands, creating potential for ambient conversations to be captured and analyzed. The timing and frequency of device usage, when lights turn on, when thermostats adjust, and when doors lock, create behavioral signatures that reveal daily routines, occupancy patterns, and lifestyle details. Internet-connected appliances, from refrigerators to televisions, generate streams of data about consumption habits, viewing preferences, and household activities that collectively paint intimate portraits of domestic life.

The Cambridge Analytica scandal exemplified how AI-driven data harvesting enables manipulation at scale. Data from millions of Facebook users, obtained through a seemingly innocuous personality quiz, was analyzed to build psychological profiles that enabled precisely targeted political advertising designed to exploit individual vulnerabilities and biases. The incident demonstrated that privacy violations affect not just individuals but democratic processes themselves, as algorithmic inference enables unprecedented capacity for behavioral manipulation.

This extraction of intimate knowledge challenges fundamental human dignity and autonomy. **1 2 3** When algorithms know individuals better than they know themselves, predicting behaviors, desires, and vulnerabilities with uncanny accuracy, people lose meaningful control over their personal information and the decisions made based upon it. The asymmetry of knowledge between individuals and the organizations deploying AI creates power imbalances that can lead to discrimination in lending, employment, insurance, and countless other domains were algorithmic inferences shape opportunities and outcomes. **1 4 2**

Understanding how AI extracts intimate knowledge from everyday data is essential for recognizing when privacy erosion threatens human dignity. The algorithms making these inferences operate according to design choices and business models that prioritize data extraction over privacy protection. **1 3** Addressing this paradox requires not just individual awareness but systemic changes to how AI systems are built, deployed, and regulated, changes that place human dignity at the center rather than treating privacy as an obstacle to algorithmic efficiency. **1 2**

Surveillance Capitalism and Behavioral Prediction: When AI Knows You Better Than You Know Yourself

The economic model that has come to dominate the digital age operates on a simple but troubling premise: human experience is raw material to be extracted, analyzed, and sold. Harvard professor Shoshana Zuboff coined the term "surveillance capitalism" to describe this system, in which companies harvest personal data not primarily to improve services for users, but to predict and influence their future behavior for profit. What distinguishes this model from earlier forms of commerce is the role of artificial intelligence in transforming mundane digital interactions into eerily accurate behavioral predictions that often surpass individuals' own self-knowledge.

The mechanics of surveillance capitalism depend on AI's capacity to detect patterns invisible to human observation. When Facebook conducted its controversial emotional contagion experiment in 2014, researchers manipulated the news feeds of nearly **700,000** users to study whether emotions could be influenced by algorithmic content curation. The study, published in the Proceedings of the National Academy of Sciences, demonstrated that users' emotional states could indeed be altered by the content they encountered, and critically, that this manipulation occurred without users' awareness or explicit consent. The experiment revealed how AI systems can identify psychological vulnerabilities and exploit them at a massive scale, fundamentally challenging assumptions about human autonomy and self-determination.

AI-driven behavioral prediction operates through continuous data extraction from every digital interaction. Search queries, purchase histories, location patterns, social media engagement, typing speed, and even the slight hesitations before clicking a link all feed machine learning models that construct detailed psychological profiles. These profiles enable what Zuboff describes as "behavioral surplus" data harvested beyond what is necessary for service improvement and instead used to predict future actions with remarkable precision. The Target pregnancy prediction algorithm that identified expectant mothers through purchase patterns represents just one application of this capability, but the same techniques extend to predicting political preferences, mental health conditions, financial stability, and relationship status.

The Cambridge Analytica scandal exposed the political dimensions of surveillance capitalism and behavioral prediction. The firm harvested data from millions of Facebook users to create psychological profiles used for targeted political advertising during the 2016 U.S. presidential election. The incident demonstrated that AI-powered behavioral prediction threatens not just individual privacy but democratic processes themselves, as algorithms enable unprecedented capacity for manipulation based on intimate knowledge of voters' fears, biases, and psychological vulnerabilities. The scandal prompted investigations across multiple countries and highlighted the inadequacy of existing regulatory frameworks to address AI-driven data exploitation.

China's social credit system represents perhaps the most comprehensive implementation of surveillance capitalism by a state actor. The system uses AI and big data to monitor citizens' behavior across multiple domains, financial transactions, social media activity, legal compliance, and even social relationships to assign scores that affect access to services, employment opportunities, and travel privileges. While Western democracies have not implemented such explicit systems, the underlying infrastructure of behavioral prediction exists throughout digital platforms, creating what privacy scholars call "soft" forms of social control through algorithmic nudging and personalized manipulation.

The ethical crisis at the heart of surveillance capitalism lies in the profound asymmetry of knowledge it creates. When AI systems know individuals better than they know themselves, predicting desires, vulnerabilities, and future actions with accuracy that exceeds human self-awareness, the foundation of autonomous choice erodes. As AI ethics researcher Kate Crawford observes, these systems are not neutral; they reflect the values and interests of their creators, who profit from behavioral prediction regardless of its impact on human dignity. The question is not whether AI can predict behavior with uncanny accuracy, but whether a society that enables such prediction can preserve the freedom and self-determination essential to human flourishing.

Facial Recognition and Biometric Surveillance: The End of Anonymity in Public Spaces

Walking through a city street once meant moving anonymously among strangers, a freedom so fundamental that few people consciously recognized it as a right. That freedom is rapidly disappearing. Facial recognition technology has transformed public spaces into zones of persistent identification, where cameras equipped with artificial intelligence can identify individuals in crowds, track their movements across locations, and build detailed records of where people go, when, and with whom. This technological shift represents more than a privacy concern; it fundamentally alters the relationship between individuals and the spaces they inhabit, threatening the anonymity that has historically protected freedom of movement, association, and expression.

The proliferation of facial recognition systems has occurred with remarkable speed and minimal public debate. Law enforcement agencies deploy technology to identify suspects, retailers use it to track customers and detect shoplifters, and transportation hubs scan travelers for security purposes. In **2020**, Robert Williams, a Black man in Michigan, was wrongfully arrested after facial recognition technology misidentified him as a shoplifting suspect. Williams spent thirty hours in detention before the error was discovered, becoming the first documented case of wrongful arrest due to facial recognition in the United States. His experience exposed not just the technology's fallibility but also its disproportionate impact on communities of color, as research consistently demonstrates that these systems misidentify people with darker skin at significantly higher rates than white individuals.

The accuracy disparities in facial recognition technology stem from biased training data and reflect broader patterns of discrimination in AI development. Joy Buolamwini's groundbreaking research at MIT revealed that commercial facial recognition systems could identify light-skinned men with near-perfect accuracy but failed to recognize dark-skinned women more than one-third of the time. These technical failures have real-world consequences when technology is deployed in law enforcement contexts, where misidentification can lead to wrongful arrests, harassment, and erosion of trust between police and communities already subject to disproportionate surveillance.

Beyond accuracy concerns, biometric surveillance systems create infrastructure for social control that can be weaponized against vulnerable populations. China's integration of facial recognition into its social credit system demonstrates how these technologies enable unprecedented state monitoring of citizens' movements, associations, and behaviors. During Hong Kong's **2019** protests, demonstrators used masks, umbrellas, and laser pointers to evade facial recognition cameras, recognizing that anonymity was essential to their ability to exercise free expression without fear of government retaliation. While Western democracies have not implemented such comprehensive surveillance systems, the underlying infrastructure exists and expands with each new deployment of facial recognition technology in public spaces.

The legal and regulatory landscape struggles to keep pace with technological capabilities. The European Union's General Data

Protection Regulation imposes strict requirements for processing biometric data, including explicit consent and purpose limitation. In the United States, no federal law specifically regulates facial recognition, though several cities, including San Francisco and Boston, have banned or restricted its use by law enforcement. This patchwork approach leaves significant gaps in protection, particularly as private companies like Clearview AI scrape billions of images from social media to build facial recognition databases used by law enforcement without individuals' knowledge or consent.

The erosion of anonymity in public spaces threatens fundamental human rights that depend on the ability to move, associate, and express oneself without constant monitoring. As privacy scholar Shoshana Zuboff observes, surveillance is not merely about watching; it is about control. When every movement through public space can be tracked, recorded, and analyzed, the freedom to participate in democratic life without fear of identification and potential retaliation diminishes. The question facing society is not whether facial recognition technology can identify individuals in crowds, but whether preserving human dignity and democratic freedoms requires protecting anonymity as a fundamental right worth defending against algorithmic surveillance.

Health Data and AI: Protecting Medical Privacy in the Age of Predictive Analytics

Medical records contain perhaps the most intimate details of human existence: diagnoses that reveal vulnerabilities, genetic information that exposes hereditary risks, mental health histories that document psychological struggles, and treatment records that chronicle the body's most private moments. When artificial intelligence systems analyze this sensitive health data to predict future medical conditions, recommend treatments, or identify at-risk populations, they create unprecedented opportunities for improved care alongside profound threats to medical privacy. The stakes extend beyond mere confidentiality; they encompass fundamental questions about who controls access to our biological and psychological information and whether individuals can maintain dignity and autonomy when algorithms know their health status better than they do themselves.

The healthcare industry has rapidly embraced AI-driven predictive analytics to transform patient care. Machine learning models now forecast which patients face elevated risks for chronic diseases like diabetes or heart failure, enabling early interventions before symptoms manifest. **75 81** These systems analyze vast datasets, electronic health records, genomic sequences, data from wearable devices, and behavioral patterns to personalize treatment plans with precision impossible through human analysis alone. **75 81** Research demonstrates that AI can predict patient deterioration, reduce hospital readmission rates, and optimize resource allocation, delivering measurable improvements in outcomes and efficiency. **75 79 82**

Yet the same characteristics that make health data valuable for AI analysis make it extraordinarily sensitive and vulnerable. Unlike credit card numbers that can be changed after a breach, genetic information and medical histories are immutable identifiers permanently linked to individuals. **79 80** The aggregation of health data from multiple sources amplifies both its analytical power and its privacy risks. Studies have shown that supposedly anonymized health datasets can often be re-identified when combined with other information, exposing patients to potential discrimination, stigmatization, and unauthorized disclosure of conditions they intended to keep private. **77 79**

The regulatory landscape attempts to balance innovation with protection through frameworks like the Health Insurance Portability and Accountability Act in the United States and the General Data Protection Regulation in Europe. **75 76 79 80 82** These laws establish strict requirements for how healthcare organizations handle, access, and share protected health information. However, the rapid evolution of AI capabilities frequently outpaces regulatory frameworks designed for earlier technologies. When AI systems can infer sensitive health conditions from seemingly innocuous data, predicting Parkinson's disease from typing patterns or detecting depression from social media activity, traditional consent models and privacy protections prove inadequate.

Johns Hopkins University implemented an AI-powered privacy analytics system that monitors every access point to patient data in real time, detecting potential violations or breaches before they escalate. **78** This initiative, which combined supervised and unsupervised machine learning with transparent analytical methods, demonstrated how healthcare institutions can proactively protect privacy while maintaining operational efficiency. The system enabled interdisciplinary collaboration and eliminated data silos that previously hindered comprehensive privacy monitoring, setting a model for other organizations seeking to balance AI innovation with rigorous privacy protection. **78**

Technical safeguards offer additional protection mechanisms. Federated learning enables AI models to learn from data distributed across multiple institutions without pooling raw patient information, preserving privacy by keeping sensitive data siloed at its source. **79** Encryption protocols protect data both at rest and in transit, while sophisticated access controls ensure that only authorized personnel can view specific information. **75 79 80** De-identification techniques remove personally identifiable information before analysis, though maintaining data utility while preventing re-identification remains technically complex, particularly as AI systems grow more sophisticated at detecting patterns. **77 79**

The ethical imperative underlying medical privacy protection centers on respect for patient autonomy and human dignity. Healthcare data should serve patients' interests, not become raw material for commercial exploitation or algorithmic profiling that occurs without meaningful consent. As one scholarly review emphasized, data privacy, bias mitigation, and accountability emerge as vital considerations in the responsible implementation of AI in healthcare. 81 When algorithms make predictions about individuals' health futures, those predictions must be transparent, explainable, and subject to human oversight that prioritizes patient welfare over efficiency or profit. 75 81 Protecting medical privacy in the age of predictive analytics requires not just technical solutions but organizational cultures that treat health data as a sacred trust, recognizing that behind every data point lies a human being whose dignity depends on maintaining control over their most intimate information.

Privacy-Preserving Technologies: Federated Learning, Differential Privacy, and Technical Solutions

The erosion of privacy through AI-driven surveillance and data extraction is not inevitable. A growing arsenal of technical solutions offers pathways to harness artificial intelligence's analytical power while protecting the personal information that underpins human dignity. These privacy-preserving technologies represent more than engineering innovations; they embody a fundamental commitment to building AI systems that respect individual autonomy and resist the surveillance capitalism model that treats human experience as raw material for extraction and profit.

Federated learning stands at the forefront of privacy-preserving machine learning approaches. Unlike traditional AI development that requires pooling sensitive data in centralized repositories, federated learning enables models to learn from data distributed across multiple devices or institutions without ever collecting the raw information itself. The technique works by training local models on individual devices, smartphones, hospital servers, or financial institutions, then sharing only the model updates, not the underlying data, with a central coordinator that aggregates improvements into a global model. Google pioneered this approach with its Gboard keyboard, allowing the system to improve predictive text suggestions by learning from millions of users' typing patterns without ever uploading their actual keystrokes to company servers.

The healthcare sector has embraced federated learning to address the tension between AI's need for large, diverse datasets and patients' right to medical privacy. Hospitals can collaboratively train diagnostic algorithms for detecting diseases like diabetic retinopathy or predicting patient deterioration without sharing protected health information across institutional boundaries. Each hospital's data remains within its secure environment, yet collective intelligence from multiple institutions improves the model's accuracy and generalizability. This approach proved particularly valuable during the COVID-19 pandemic, when researchers needed to rapidly develop predictive models from patient data scattered across healthcare systems worldwide without compromising privacy protections.

Differential privacy offers complementary protection through mathematical guarantees that individual data points cannot be reverse engineered from AI system outputs. The technique adds carefully calibrated statistical noise to data or query results, ensuring that the presence or absence of any single person in a dataset cannot be determined even by sophisticated attackers. Computer scientist Cynthia Dwork, who pioneered differential privacy research at Microsoft, describes it as providing "plausible deniability" for individuals whose information contributes to aggregate analyses. The U.S. Census Bureau adopted differential privacy for the 2020 census, balancing the constitutional mandate to count every person with privacy protections against re-identification attacks that could expose individuals' sensitive demographic information.

Apple's implementation of differential privacy in iOS demonstrates how technology enables data-driven service improvements without surveillance. The company collects usage statistics about which emojis users favor, which websites frequently crash Safari, and which words trigger autocorrect frustrations, all while ensuring that no individual user's behavior can be isolated or identified. The noise injection means Apple learns population-level patterns useful for product development without creating detailed profiles of specific users' digital lives.

Additional technical safeguards complement these core approaches. Homomorphic encryption allows computation on encrypted data without decryption, enabling organizations to outsource AI processing to cloud providers without exposing sensitive information. Secure multi-party computation permits multiple entities to jointly analyze data while keeping each party's inputs private. Privacy-preserving data anonymization techniques, such as k-anonymity, ensure that individuals cannot be identified in datasets released for research or analysis.

These technologies face real-world challenges. Federated learning requires careful defense against model poisoning attacks where malicious participants corrupt the global model. Differential privacy involves trade-offs between privacy guarantees and analytical accuracy; more noise provides stronger protection but reduces data utility. Implementation complexity and computational costs can discourage adoption, particularly among smaller organizations lacking technical expertise.

Yet the existence and continued development of privacy-preserving technologies demonstrate that the choice between AI innovation and privacy protection is false. Technical solutions exist that honor both imperatives, if organizations and policymakers prioritize human dignity over convenience and profit. The question is not whether privacy-preserving AI is possible, but whether society will demand and implement it.

Regulatory Frameworks and Individual Rights: GDPR, Data Sovereignty, and the Future of Privacy Protection

The teenager whose pregnancy Target predicted through algorithmic analysis of her shopping patterns had no meaningful way to prevent that inference, challenge its accuracy, or control how the company used that knowledge. Her experience, and millions like it, exposed a fundamental gap in privacy protection: traditional frameworks built around notice and consent proved inadequate when AI systems could extract sensitive information from seemingly innocuous data without explicit disclosure. **2** This inadequacy catalyzed a global rethinking of privacy rights, culminating in regulatory frameworks that attempt to restore individual control over personal information in the age of artificial intelligence.

The European Union's General Data Protection Regulation, implemented in **2018**, represents the most comprehensive attempt to establish privacy rights that address AI-specific challenges. **2. 67** GDPR grants individuals several critical rights that directly constrain how organizations can deploy AI systems. The right to access permits individuals to obtain confirmation of whether their personal data is being processed and to receive a copy of that data in a commonly used electronic format, creating accountability mechanisms that force organizations to maintain transparency about what information they hold and how it is being used. The right to erasure, commonly called the right to forget, allows individuals to request deletion of their personal data under certain circumstances, a principle that poses challenges for machine learning models that cannot easily forget data without complete retraining, yet reflects the fundamental assertion that individuals should not be permanently defined by historical information.

Perhaps most significantly for AI systems, GDPR establishes a right to explanation for automated decision-making that produces legal or similarly significant effects. **67** When an algorithm denies someone a loan, rejects a job application, or determines insurance premiums, individuals have the right to understand the logic behind that decision. This requirement directly challenges the opacity of complex machine learning systems, particularly deep learning models where even developers cannot fully articulate why a model reached a particular conclusion. 67 Organizations deploying AI must now balance algorithmic sophistication with explainability, often requiring simplified models or interpretability tools that can provide meaningful explanations to affected individuals.

GDPR also introduced Data Protection Impact Assessments, requiring organizations to evaluate privacy and security risks before deploying new technologies. For AI systems, these assessments serve as critical checkpoints where potential harm can be identified and mitigated before deployment at scale. The regulation mandates the appointment of Data Protection Officers in certain organizations, creating dedicated roles responsible for ensuring compliance and serving as points of contact for regulatory authorities.

The concept of data sovereignty, the principle that data is subject to the laws of the nation where it is collected and that individuals and nations should maintain control over their information assets, has emerged alongside regulatory frameworks like GDPR. 2 Different nations have adopted varying approaches: China emphasizes localization requirements mandating that certain data categories remain within Chinese territory, while India balances data access for innovation with localization requirements for sensitive information. Brazil's Lei Geral de Protea de Dados mirrors GDPR while adding specific provisions for data localization in critical sectors.

These regulatory frameworks create inherent tensions with AI development's global nature. Machine learning models often require large, diverse datasets to function effectively. Strict localization requirements can fragment datasets across jurisdictions, potentially reducing model quality while increasing costs for organizations operating across multiple regions. Yet unrestricted data flows to centralized processing centers raise legitimate concerns about surveillance, loss of national control over strategic information, and the concentration of power among multinational technology companies.

The future of privacy protection will likely involve continued evolution of regulatory frameworks as AI capabilities advance, alongside technical solutions like differential privacy and federated learning that enable beneficial data analysis while protecting individual rights. The question is not whether privacy can survive the age of AI, but whether societies will demand and enforce protections that place human dignity above algorithmic efficiency. 2 The teenager whose pregnancy Target predicted through shopping patterns, the wrongfully arrested man misidentified by facial recognition, and the patients whose medical records feed predictive algorithms, these cases reveal a fundamental truth about privacy in the age of artificial intelligence. The erosion of personal privacy is not an inevitable consequence of technological progress but the result of deliberate choices about how AI systems are designed, deployed, and governed. Every algorithm that extracts intimate knowledge from everyday data, every surveillance system that tracks faces in crowds, and every predictive model that analyzes health information reflects human decisions about whose interests matter most: the individuals whose lives are exposed to or the organizations that profit from exposure.

The privacy paradox that defines our current moment stems from this asymmetry of power and knowledge. Individuals surrender intimate details through digital interactions they perceive as mundane, while sophisticated AI systems detect patterns invisible to human awareness and make inferences that surpass people's own self-knowledge. This imbalance threatens more than confidentiality; it undermines the autonomy and self-determination essential to human dignity. When algorithms know individuals better than they know themselves, when every movement through public space can be tracked and recorded, and when medical conditions can be predicted from typing patterns, the foundation of authentic choice erodes. People cannot meaningfully consent to uses of their information they cannot anticipate or understand, particularly when the extraction happens invisibly through inferential analytics that reveal what was never explicitly disclosed.

Yet the existence of privacy-preserving technologies and emerging regulatory frameworks demonstrates that different futures remain possible. Federated learning enables AI systems to learn from distributed data without centralized collection. Differential privacy provides mathematical guarantees against re-identification. The European Union's General Data Protection Regulation establishes rights to access, erasure, and explanation that restore some measure of individual control. These technical and legal innovations prove that the choice between AI's benefits and privacy protection is false, both can coexist when human dignity takes precedence over algorithmic efficiency and corporate profit.

The challenges ahead require sustained attention across multiple domains. Technical solutions must continue evolving to address AI capabilities that advance faster than safeguards. Regulatory frameworks need refinement as new applications emerge that existing laws never anticipated. Organizations deploying AI systems must prioritize privacy protection not as a compliance obligation but as an ethical imperative. Individuals need education and tools to understand how their data is collected, analyzed, and used to make decisions affecting their lives. Most fundamentally, societies must decide whether to accept surveillance capitalism's premise that human experience exists as raw material for extraction or to demand economic models that respect privacy as a human right worth protecting even when violation would be profitable.

The stakes extend beyond individual privacy violations to encompass threats to democratic participation, freedom of expression, and the capacity for authentic human connection in spaces free from constant monitoring. Anonymity in public spaces, confidentiality in medical contexts, and the ability to explore ideas without surveillance are not luxuries but prerequisites for flourishing in free societies. AI systems that erode these protections in pursuit of behavioral prediction and control represent choices, not inevitabilities. These choices can be challenged, regulated, and changed by people who recognize that technology must serve human dignity rather than diminish it.

The future of privacy in the AI age depends on whether individuals, organizations, and governments act now to establish boundaries that preserve what makes us fundamentally human. The algorithms analyzing our data, tracking our movements, and predicting our futures operate according to values embedded by their creators. The question is not whether AI will respect privacy, but whether we will demand and build systems that do. That choice remains ours to make, but the window for making it narrows with each new deployment of surveillance infrastructure and each normalization of algorithmic intrusion into intimate life. Privacy protection requires not passive acceptance but active resistance to the premise that efficiency and profit justify the sacrifice of human dignity.

CHAPTER 5: AI GOVERNANCE FRAMEWORK AND ACCOUNTABILITY STANDARDS: WHO DECIDES, WHO ANSWERS?

When Boeing's 737 MAX aircraft crashed twice in 2019, killing 346 people, investigators traced the disasters to flawed software that pilots couldn't override, but determining who should be held accountable proved far more complex than identifying the technical failure. The crashes exposed a governance vacuum: aviation regulators had delegated safety assessments to the manufacturer, software developers had made critical design decisions without full oversight, and no clear framework existed for assigning responsibility when AI-assisted systems failed with catastrophic consequences. The tragedy revealed a fundamental truth about artificial intelligence in high-stakes systems: technical sophistication means nothing without clear governance structures and accountability mechanisms. As AI systems become more autonomous, more consequential, and more deeply embedded in critical infrastructure, the question of who decides how they operate and who answers when they fail has become one of the most urgent challenges of our time.

The governance vacuum surrounding AI is not an accident; it reflects the unprecedented speed at which these technologies have evolved, and the difficulty traditional regulatory institutions face in keeping pace with innovation. While lawmakers debate legislation, AI systems are already making decisions about who receives medical treatment, who gets approved for loans, who is flagged as security risk, and which neighborhoods receive police attention. These decisions have profound consequences for human lives, yet the frameworks for oversight, accountability, and redress remain fragmented, inconsistent, and often nonexistent.

The challenge extends beyond simply creating rules. Effective AI governance must navigate a complex landscape of competing interests and legitimate tensions. Technology companies argue that excessive regulation will stifle innovation and hand competitive advantages to nations with a lighter regulatory touch. Civil society advocates warn that without strong oversight; AI will deepen existing inequalities and concentrate power in the hands of a few corporations. Governments struggle to regulate technologies they barely understand, often relying on the very companies they seek to oversee for technical expertise. Meanwhile, AI systems continue to evolve and proliferate faster than any single institution can monitor or control.

Different regions have adopted strikingly different approaches to this governance challenge, each reflecting distinct cultural values and political philosophies. The European Union has moved toward comprehensive regulation with its AI Act, categorizing systems by risk level and imposing strict requirements on high-risk applications. The United States has favored a more fragmented, sector-specific approach, with different agencies developing guidelines for AI in their domains. China has implemented a model that emphasizes government control and alignment with national interests. Each approach offers insights and limitations, and none has yet proven definitively successful in balancing innovation with protection.

Beyond government regulation, questions of corporate governance and self-regulation loom large. When companies develop and deploy AI systems, who within those organizations bears responsibility for ethical outcomes? How should boards of directors oversee AI risks they may not fully understand? What role should ethics committees, impact assessments, and internal accountability mechanisms play? The answers matter enormously, because in practice, many of the most consequential decisions about AI are made not by regulators or legislators but by engineers, product managers, and executives working within corporate structures that may prioritize growth and efficiency over caution and equity.

The accountability question becomes even more complex when AI systems cause harm. Traditional legal frameworks assume clear chains of causation and identifiable responsible parties, concepts that blur when autonomous systems make decisions based on training data compiled by one team, algorithms designed by another, and deployment choices made by a third party. If a medical AI recommends a treatment that harms a patient, who is liable? The hospital that deployed it? The company that sold it? The developers who built it? The data scientists who trained it? Current legal doctrines struggle to provide clear answers, leaving victims without recourse and creating uncertainty that paradoxically may discourage responsible practices.

This chapter examines these governance challenges not as abstract policy debates but as practical questions that affect how AI shapes our lives, our work, and our society. Understanding who holds power over AI systems and who bears responsibility when they fail is essential for anyone seeking to navigate or influence the AI-transformed world we are creating.

The Accountability Gap: Why Traditional Legal Frameworks Fail for AI Systems

Traditional legal systems operate on a fundamental assumption: when harm occurs, responsibility can be traced to identifiable human actors or organizations through clear chains of causation. A surgeon makes an error during an operation. A manufacturer produces a defective product. A driver runs a red light. In each case, existing legal frameworks, negligence, product liability, and tort law provide mechanisms for determining faults and assigning accountability. **87** But artificial intelligence systems shatter this assumption, creating what legal scholars call the "accountability gap," a space where harm occurs yet no party can be held fully responsible under existing legal doctrines. 83 84 87

The problem begins with the technical nature of AI systems themselves. Modern machine learning models, particularly deep neural networks, function as what researchers describe as "black boxes," systems so complex that even their creators cannot fully explain how they arrive at specific decisions. 83 86 When an AI system denies someone a loan, recommends a medical treatment, or flags an individual as a security risk, tracing that decision back to a specific line of code or design choice often proves impossible. 83 86 The decision emerges from millions of weighted connections shaped by training data, algorithmic architecture, and learning processes that no single person fully controls or comprehends.

This opacity creates what legal scholars identify as a diffusion of responsibility across multiple parties. 86 The data scientists who compiled the training dataset claim they merely gathered information. The engineers who designed the algorithm argue they created a neutral tool. The company that deployed the system points to the developers. The executives who approved the purchase cite operational teams. 86 Each party can plausibly deny full responsibility, yet collectively they have created a system that makes consequential decisions affecting human lives. Traditional legal frameworks, built to assign clear liability, struggle to navigate this distributed accountability. 86

Product liability law illustrates the challenge. These doctrines typically address physical defects or clear malfunctions, a medication that causes unexpected harm. 87 But AI systems rarely "malfunction" in the traditional sense. Instead, they produce probabilistic outputs based on patterns in data, making decisions that may be statistically reasonable yet individually harmful or discriminatory. When an AI system denies healthcare coverage based on subtle patterns in patient data, has the product malfunctioned, or has it simply revealed biases embedded in historical medical practices? 87 Current product liability frameworks provide no clear answer.

Negligence standards face similar limitations. 87 Proving negligence requires demonstrating that a party failed to exercise reasonable care, but what constitutes reasonable care when deploying AI systems remains undefined. Should organizations be required to understand every decision their AI makes? To audit for bias in ways that may be technically impossible. To maintain human oversight that defeats efficiency gains AI promises. Courts have offered inconsistent guidance, leaving organizations uncertain about their legal obligations and victims uncertain about their rights to redress.

The accountability gap has real consequences documented in court systems worldwide. Over 120 judicial decisions have involved AI errors, from fabricated legal citations generated by language models to faulty risk assessments in criminal justice. 85 In June 2023, New York lawyers faced sanctions after submitting legal briefs citing six entirely fictitious court cases created by ChatGPT, an incident that demonstrated how organizations deploying AI without verification mechanisms face liability despite claiming ignorance of the system's limitations. 85

Perhaps most troubling, governments increasingly disclaim responsibility for AI-assisted decisions by citing the complexity of third-party systems they deploy. 84 When challenged on welfare determinations or criminal justice outcomes influenced by algorithmic tools, officials argue they cannot be held accountable for systems they do not fully understand. 84 This creates what researchers describe as "algorithmic systems contributing to government decision-making without mechanisms of accountability or liability." 84

Legal systems have consistently rejected the notion that "the AI did it" absolves human responsibility. 85 Courts maintain that ultimate accountability remains with the humans and organizations that deploy, oversee, or benefit from AI systems. 85 Yet this principle offers little practical guidance for assigning responsibility in complex sociotechnical systems where decisions emerge from interactions between data, algorithms, and human choices made by dozens of parties across organizational and jurisdictional boundaries.

Regulatory Approaches Across Jurisdictions: EU AI Act, U.S. Sectoral Regulation, and China's Governance Model

The absence of a unified global approach to AI governance has produced three distinct regulatory models, each reflecting different values about the relationship between technology, individual rights, and state power. 1, 2, 3 The European Union, United States, and China have emerged as the world's primary regulatory laboratories, developing frameworks that will likely shape AI governance worldwide for decades to come. Understanding these approaches is essential for anyone seeking to navigate the complex landscape of AI accountability, whether as a developer, user, policymaker, or citizen affected by these systems. 1

The European Union has taken the most comprehensive approach with its AI Act, formally adopted in 2024 after years of deliberation. 2 This landmark legislation establishes a risk-based framework that categorizes AI systems according to the potential harm they pose. 2 At the strictest level, certain applications are banned outright, including social scoring systems by governments and real-time biometric identification in public spaces except under narrow exceptions. High-risk AI systems, those used in critical infrastructure, education, employment, law enforcement, and essential services, face stringent requirements, including conformity assessments, transparency obligations, data governance standards, human oversight mechanisms, and continuous post-market monitoring.

The EU approach embeds accountability at multiple levels. 2 Providers of high-risk AI systems must demonstrate compliance before deployment, maintain detailed documentation, and ensure their systems meet accuracy and robustness standards. Deployers, the organizations using these systems, bear responsibility for proper implementation and monitoring. National supervisory authorities enforce compliance, while a European Artificial Intelligence Board coordinates oversight across member states. Violations can result in substantial fines reaching up to seven percent of global annual revenue for the most serious infractions. This framework reflects European values prioritizing fundamental rights, transparency, and precautionary principles even when they constrain innovation speed. 2

The United States has adopted a strikingly different path, favoring sectoral regulation through existing agencies rather than comprehensive legislation. 2 The Food and Drug Administration oversees AI-enabled medical devices, the Federal Trade Commission addresses AI in consumer protection and privacy contexts, the Equal Employment Opportunity Commission enforces anti-discrimination laws that apply to AI hiring tools, and the National Highway Traffic Safety Administration regulates autonomous vehicles. This fragmented approach allows specialized agencies to apply domain expertise but creates inconsistency across sectors and potential gaps where AI applications fall between regulatory jurisdictions.

Rather than mandatory requirements, U.S. governance emphasizes voluntary frameworks and guidance. The National Institute of Standards and Technology has developed an AI Risk Management Framework that organizations can adopt voluntarily, focusing on trustworthiness characteristics including validity, reliability, safety, security, and fairness. Recent executive orders have directed agencies to ensure safe

AI development, but these lack the binding force of legislation. Some states have begun filling the federal vacuum; California, for instance, has enacted laws addressing automated decision-making and algorithmic discrimination, creating a patchwork of requirements that companies must navigate. This approach reflects American preferences for innovation, flexibility, market-driven solutions, and skepticism toward comprehensive regulation, though critics argue it leaves significant accountability gaps.

China has implemented a centralized, state-led governance model that treats AI as a strategic technology requiring direct government control. 2 The country's New Generation Artificial Intelligence Development Plan, announced in 2017, set ambitious targets for AI leadership by 2030 while establishing clear state oversight. Subsequent regulations have addressed specific AI applications: the Algorithmic Recommendation Provisions require transparency from platforms using recommendation algorithms, while Deep Synthesis Regulations mandate labeling of AI-generated content, including deepfakes. The Cyberspace Administration of China enforces these rules with authority to impose penalties, suspend services, and require algorithmic modifications.

China's model emphasizes alignment with state priorities, including social stability, national security, and ideological control alongside innovation goals. AI systems must undergo security assessments, and companies face obligations to prevent content that undermines state authority or social order. This approach enables rapid, coordinated implementation of governance standards but raises concerns about surveillance, censorship, and the subordination of individual rights to state interests. The model demonstrates how AI governance inevitably reflects broader political values about the proper relationship between individuals, technology companies, and government power.

These three models present fundamentally different answers to the question of who decides how AI operates and who answers when it fails, with profound implications for innovation, rights protection, and the future distribution of power in an AI-shaped world. 1, 2, 3

Corporate Governance and Self-Regulation: When Industry Sets Its Own Rules

Between government regulation and complete autonomy lies a middle ground where corporations establish their own rules for AI development and deployment. This self-regulatory approach has become increasingly prominent as technology companies argue they possess the technical expertise and agility to govern AI more effectively than traditional regulatory bodies. 89 Yet this arrangement raises a fundamental question: can organizations be trusted to police themselves when ethical considerations may conflict with competitive advantage and profit motives? 89, 91

Corporate self-regulation in AI typically manifests through internal governance frameworks that establish principles, processes, and oversight structures. IBM's widely cited AI ethics governance model exemplifies this approach, featuring a multilayered system that includes a Policy Advisory Committee setting global objectives, an AI Ethics Board defining and maintaining ethics policies, AI Ethics Focal Points within business units who identify risks, and an advocacy network that fosters ethical culture throughout the organization. 88 This structure enables what IBM describes as "AI ethics review at scale," with focal points driving compliance and initiating reviews before high-risk projects proceed. 88

The core principles underlying corporate AI governance frameworks show remarkable consistency across industry leaders. Fairness, transparency, accountability, privacy, and security appear repeatedly in the ethical guidelines published by major technology companies. 90, 91 These principles are operationalized through concrete processes: assessing intended use and potential harms, conducting regulatory compliance checks, reviewing systems for bias and discrimination, assigning clear responsibilities for oversight, and providing ongoing employee education. **88, 89** When functioning as intended, these frameworks can identify problems before deployment, as occurred when an IBM business unit's AI hiring project was flagged for potential discrimination, leading to algorithmic retraining and transparency improvements before release. **88**

The advantages of industry self-regulation are significant. Corporations possess deep technical expertise that enables a nuanced understanding of AI capabilities and limitations. **89 They** can adapt governance approaches rapidly as technologies evolve, avoiding the delays inherent in legislative processes. **89 structures** can be embedded throughout large organizations, creating scalable oversight mechanisms that reach individual projects and teams. **88** Flexibility has allowed some companies to establish governance practices that exceed existing legal requirements, setting standards that may eventually inform formal regulation.

Yet the limitations and risks of self-regulation are equally substantial. The fundamental tension between profit maximization and ethical constraint creates inherent conflicts of interest. **89, 91,** Without external oversight, enforcement mechanisms remain vulnerable to internal pressure to prioritize business objectives over ethical concerns. **89, 91** The **2018** controversy surrounding Google's involvement in Project Maven, a military AI initiative that violated the company's stated principles until employee protests forced withdrawal, demonstrated how corporate commitments can bend under competitive or financial pressure. **89** Self-regulatory frameworks lack the independence and enforcement power of external regulation, making them susceptible to what critics describe as "ethics washing", the appearance of robust governance without meaningful constraint on harmful practices. **89, 91**

The inconsistency across corporate approaches creates additional challenges. 91 While some companies have developed detailed governance structures, others offer only vague principles with minimal implementation detail. This fragmentation leaves gaps in protection and creates confusion about industry standards. Public trust remains limited; surveys consistently show skepticism about whether technology companies will self-regulate effectively without external accountability mechanisms. 89, 91

The reality emerging from corporate self-regulation efforts suggests that internal governance frameworks serve an important but insufficient role. They can foster ethical awareness, create processes for identifying risks, and enable rapid response to emerging challenges. 89 However, they cannot replace independent oversight, legal accountability, or democratic input into decisions that affect millions of people. 89, 91, The question is not whether corporations should establish internal governance, they must, but whether self-regulation alone can ensure AI systems serve human welfare rather than merely corporate interests.

Multi-Stakeholder Governance: Balancing Government, Industry, Civil Society, and Technical Expertise

Neither government regulation alone nor corporate self-governance can adequately address the complex challenges AI presents. Technology's rapid evolution, global reach, and profound societal impacts demand a more inclusive approach, one that brings diverse stakeholders to the table with genuine authority to shape outcomes. Multi-stakeholder governance has emerged as a promising framework for balancing the technical expertise of industry, the democratic legitimacy of government, the advocacy power of civil society, and the analytical rigor of academic institutions.

The rationale for multi-stakeholder approaches rests on a simple recognition: no single entity possesses all the knowledge, authority, and perspective needed to govern AI effectively. Governments bring regulatory power and democratic accountability but often lack the technical expertise to understand rapidly evolving systems. Technology companies possess deep technical knowledge and implementation capacity but face inherent conflicts between profit motives and public welfare. Civil society organizations represent affected communities and advocate for rights but may lack resources and technical understanding. Academic researchers offer independent analysis and ethical frameworks but operate outside implementation contexts. Each stakeholder brings essential yet incomplete contributions to the governance challenge.

International initiatives have begun operating this collaborative model. The Global Partnership on AI, launched in **2020**, brings together countries, businesses, and researchers to develop shared principles and best practices for responsible AI development. The World Economic Forum's AI Governance Alliance convenes over two hundred leaders from industry, academia, civil society, and government to guide AI development through collective deliberation. The Organization for Economic Co-operation and Development's AI principles, developed through extensive multi-stakeholder consultation, have influenced national policies worldwide. These initiatives demonstrate that diverse groups can reach meaningful consensus on governance standards despite different interests and perspectives.

The practical benefits of multi-stakeholder governance extend beyond legitimacy to effectiveness. When healthcare AI governance frameworks incorporate input from patients, clinicians, technologists, ethicists, and regulators, the resulting standards address a broader range of concerns than any single group would identify. A **2023** study on AI governance in healthcare proposed establishing a Health AI Consumer Consortium specifically to ensure patient-centered oversight alongside technical and regulatory perspectives. This inclusive approach helps identify risks that technical experts might overlook, implementation challenges that policymakers might not anticipate, and equity concerns that industry might not prioritize.

Yet multi-stakeholder governance faces significant challenges that cannot be dismissed. Power imbalances frequently undermine the ideal of equal participation. Large technology companies and well-resourced governments can dominate discussions through superior access to information, technical expertise, and negotiating capacity. Civil society representatives from marginalized communities or developing nations often lack the resources to participate meaningfully in lengthy deliberative processes. The result can be governance frameworks that claim multi-stakeholder legitimacy while reflecting the preferences of the most powerful participants.

The complexity and slowness of consensus-building processes present additional obstacles. When urgent decisions are needed, as with rapidly deployed generative AI systems, the time required for genuine multi-stakeholder deliberation may seem incompatible with the pace of technological change. Coordination across diverse groups with different priorities, vocabularies, and institutional cultures requires substantial effort and patience. Some critics argue that multi-stakeholder processes can become forums for endless discussion that delay necessary action.

Accountability presents perhaps the most fundamental challenge. When responsibility is shared across multiple stakeholders, determining who answers for failures becomes difficult. If a governance framework developed through multi-stakeholder collaboration fails to prevent harm, which participants bear responsibility? Shared governance can inadvertently create the same accountability gaps that plague AI systems themselves, a diffusion of responsibility that leaves victims without clear recourse.

Despite these limitations, the alternative approaches have proven inadequate. Top-down government regulation struggles with technical complexity and innovation speed. Industry self-regulation lacks independence and enforcement power. Multi-stakeholder governance, for all its imperfections, offers the most promising path toward AI systems that reflect diverse values, serve broad interests, and maintain democratic accountability while harnessing technical expertise. The challenge lies not in whether to pursue multi-stakeholder approaches but in designing them to overcome power imbalances, maintain efficiency, and preserve clear lines of accountability even as authority is shared.

Liability and Legal Responsibility: Who Pays When AI Systems Cause Harm?

When a self-driving car strikes a pedestrian, when a medical AI recommends a treatment that harms a patient, when an algorithmic hiring tool systematically excludes qualified candidates, who bears legal responsibility? Traditional legal frameworks assume clear chains of causation linking harm to identifiable parties, but AI systems shatter this assumption. The question of liability has become one of the most urgent challenges in AI governance, with profound implications for victims seeking justice, companies deploying these systems, and society's ability to ensure accountability.

The complexity begins with AI's multi-party ecosystem. Unlike traditional products with clear manufacturers, AI systems involve developers who create algorithms, data providers who supply training information, deployers who implement systems in specific contexts, and users who interact with outputs. **92 Each** party exercises different degrees of control and knowledge over system behavior. **92** When harm occurs, determining which party should bear responsibility, and to what degree, presents challenges that existing legal doctrines struggle to address.

Traditional negligence law requires proving that a defendant owed a duty of care and failed to exercise reasonable caution. **94** Yet establishing this duty becomes problematic when developers lose control over AI systems after deployment, particularly through licensing agreements or cloud-based services. 94 The opacity of AI decision-making compounds the difficulty. Modern machine learning models, especially deep neural networks, function as "black boxes" where even creators cannot fully explain specific outputs. 93 creates unique evidentiary burdens for claimants attempting to prove that an AI system caused their harm.

Product liability law offers an alternative approach by holding manufacturers strictly liable for defective products regardless of negligence. Historically, however, courts found that software not embedded in hardware did not constitute a "product" under these frameworks, creating significant protection gaps. 94 This legal landscape is shifting dramatically. 92 The European Union's New Product Liability Directive, effective August 2026, explicitly extends strict liability to software and AI systems, whether embedded in hardware or distributed independently. 94 The directive presumes defectiveness in cases involving complex AI systems where claimants face excessive difficulties proving their case, fundamentally shifting the burden of proof to defendant companies. 94

In the United States, proposed legislation like the AI LEAD Act represents a similar evolution. The bill explicitly classifies AI systems as products subject to product liability claims based on design defects, failure to warn, and unreasonably dangerous characteristics. 92 Critically, it extends liability beyond developers to deployers who substantially modify or intentionally misuse AI systems, acknowledging that responsibility should not rest solely with original creators. 92 This approach attempts to distribute accountability proportionally among parties based on their ability to identify and mitigate risks. 93

Some jurisdictions are experimenting with novel frameworks that depart from traditional tort law. Rhode Island's pending legislation establishes a rebuttable presumption that if an AI caused harm requiring examination of mental state, the system possessed the requisite intent to be held liable. 95 This presumption significantly eases plaintiffs' burden given the opacity of AI models and the challenge of proving intent 95.

The practical implications are substantial. Organizations developing and deploying AI systems face increasing requirements for rigorous safety assessments, transparent documentation of decision-making processes, and clear warnings about limitations and potential harms. 92 Contractual attempts to waive liability in unreasonable ways are increasingly disfavored by courts and regulators. 92 The New Product Liability Directive explicitly prevents software developers from excluding liability for death or personal injury, even when supplying components to larger manufacturers. 94

Yet significant uncertainties remain. How should causation be assessed when AI systems operate within complex socio-technical environments where humans make final decisions? Should developers bear liability for emergent properties they did not anticipate? How will international commerce function as different jurisdictions develop divergent liability frameworks? These questions will shape not only legal doctrine but also the fundamental incentives governing how AI systems are designed, tested, and deployed in contexts affecting human lives.

International Coordination and Global Standards: Governing Technology That Knows No Borders

Artificial intelligence recognizes no borders. An algorithm trained in California can screen job applicants in Tokyo, recommend medical treatments in Berlin, and determine creditworthiness in São Paulo, often within milliseconds. This borderless nature creates a fundamental governance challenge: how can individual nations regulate technologies that operate globally, learn from international data, and affect populations far beyond any single jurisdiction's reach? The answer increasingly points toward international coordination and harmonized global standards, yet achieving such coordination has proven extraordinarily difficult.

The need for global AI governance stems from practical realities that transcend political boundaries. When a facial recognition system trained on predominantly Western faces performs poorly on Asian or African populations, the bias affects people worldwide. When social media algorithms amplify misinformation, democratic processes suffer across continents. When autonomous weapons systems are developed without international consensus on acceptable use, global security deteriorates. These challenges cannot be solved by any nation acting alone, no matter how comprehensive its domestic regulations.

Several international initiatives have emerged to address this coordination imperative. The Organization for Economic Co-operation and Development's AI Principles, adopted in 2019 and endorsed by over forty countries, represent the first intergovernmental standard on AI. These principles promote AI systems that are innovative, trustworthy, and respectful of human rights and democratic values. While non-binding, they have influenced national policies worldwide and established common language for international dialogue. The UNESCO Recommendation on the Ethics of Artificial Intelligence, adopted in 2021 by all 193 member states, went further by creating the first global standard-setting instrument specifically addressing AI ethics. This framework emphasizes human dignity, fairness, transparency, and accountability while calling on member states to implement aligned national policies.

The Global Partnership on Artificial Intelligence, launched in 2020, takes a different approach by bringing together governments, industry, civil society, and academia to bridge the gap between AI principles and practice. Through working groups focused on responsible AI, data governance, and societal impacts, GPAI develops practical tools and guidance that member countries can adapt to their contexts. These initiatives demonstrate that international consensus on broad principles is achievable even amid geopolitical tensions.

Yet translating principles into enforceable standards reveals profound challenges. Nations approach AI governance through fundamentally different cultural, legal, and political lenses. The European Union's comprehensive AI Act emphasizes precautionary principles and fundamental rights protection. The United States favors sectoral regulation and market-driven innovation. China implements centralized state oversight aligned with national strategic priorities. These divergent approaches reflect deeper disagreements about the proper balance between innovation and regulation, individual rights and collective interests, and market freedom and government control.

Geopolitical competition further complicates coordination efforts. When AI is viewed as a strategic technology critical to economic and military advantage, nations become reluctant to accept international standards that might constrain their capabilities or advantage competitors. The rapid pace of AI development exacerbates this challenge; by the time international bodies reach consensus on standards for one generation of technology, the next has already emerged with novel capabilities and risks.

Enforcement presents perhaps the most fundamental obstacle. International agreements typically lack binding enforcement mechanisms, relying instead on voluntary compliance and reputational pressure. When violations occur, no global authority can impose meaningful consequences. This weakness becomes particularly problematic when addressing AI systems deployed by powerful corporations or governments with limited accountability to international norms.

Despite these formidable barriers, the alternative to international coordination is a fragmented regulatory landscape that serves no one well. Companies face compliance burdens navigating inconsistent national requirements. Citizens receive uneven protection depending on their location. The race to the bottom becomes a real risk as jurisdictions compete to attract AI investment through lax oversight. The question is not whether international coordination is necessary; it clearly is, but whether the global community possesses the political will to overcome national interests and achieve it. The Boeing 737 MAX crashes that opened this chapter exposed more than a technical failure; they revealed the profound inadequacy of existing governance structures to address AI-assisted systems in high-stakes contexts. The tragedy demonstrated that sophisticated technology without clear accountability mechanisms creates not innovation but danger. As artificial intelligence becomes increasingly autonomous and consequential, the question of who decides how these systems operates and who answers when they fail has moved from theoretical debate to urgent necessity. The governance frameworks established today will determine whether AI amplifies human dignity or undermines it, whether these systems serve the many or concentrate power among the few.

The accountability gap remains the central challenge confronting AI governance. Traditional legal frameworks built on assumptions of clear causation and identifiable responsible parties struggle to address systems where decisions emerge from complex interactions between data, algorithms, and human choices distributed across organizational and jurisdictional boundaries. When AI systems cause harm, victims face evidentiary burdens that existing tort law was never designed to handle, while defendants exploit opacity and distribute responsibility to avoid liability. The evolution of product liability frameworks in the European Union and proposed legislation in the United States represent significant steps toward closing this gap, yet fundamental questions about causation, foreseeability, and proportional responsibility remain unresolved.

The divergent regulatory approaches across major jurisdictions reflect deeper disagreements about values, priorities, and the proper relationship between innovation and precaution. The European Union's comprehensive, rights-based framework stands in stark contrast to America's fragmented sectoral approach and China's centralized state control model. Each approach offers insights and limitations, and none has yet proven definitively successful in balancing the competing demands of fostering innovation, protecting rights, ensuring security, and maintaining democratic accountability. These differences will shape not only how AI develops within each region but also the global landscape as companies navigate inconsistent requirements and nations compete for technological advantage.

Corporate self-regulation and multi-stakeholder governance represent attempts to fill the vacuum left by inadequate government oversight, yet both approaches face inherent limitations. Internal governance frameworks can foster ethical awareness and create processes for identifying risks, but they cannot replace independent oversight or legal accountability when profit motives conflict with public welfare. Multi-stakeholder initiatives bring diverse perspectives and technical expertise to governance challenges, but power imbalances, coordination difficulties, and diffused accountability threaten to undermine their effectiveness. Neither approach eliminates the need for robust government regulation with meaningful enforcement mechanisms.

The borderless nature of AI technology demands international coordination that has proven extraordinarily difficult to achieve. While organizations like the OECD and UNESCO have established shared principles, translating consensus on broad values into enforceable standards requires overcoming geopolitical competition, divergent legal traditions, and the absence of binding enforcement mechanisms. Yet the alternative, a fragmented regulatory landscape where standards vary dramatically across jurisdictions, serves neither innovation nor protection. The race to develop global governance frameworks is not merely a policy challenge but a test of whether humanity can coordinate collective action on technologies that recognize no borders.

The governance decisions made in the coming years will reverberate for decades. They will determine whether AI systems operate with transparency or opacity, whether accountability mechanisms protect victims or shield powerful actors, and whether development prioritizes human welfare or corporate advantage. These are not questions that can be delegated to technical experts or left to market forces; they are fundamentally democratic choices about the kind of society we want to create. The frameworks established now will shape whether artificial intelligence becomes a tool for human flourishing or a mechanism for concentrating power and deepening inequality. The window for making these choices deliberately rather than by default is closing, but it has not yet closed.

CHAPTER 6: PRACTICAL AI ETHICS IN MEDICINE, LONGEVITY, AND EVERYDAY LIFE

When a radiologist studies a patient's brain scan, they don't work alone; an AI system can flag subtle abnormalities that might have escaped human eyes during a busy shift. As someone living with Parkinson's disease for over two decades and a former trauma nurse who has witnessed both the miracles and limitations of medical technology, I understand that this collaboration between human expertise and artificial intelligence represents both extraordinary promise and profound ethical responsibility. This intersection of human expertise and machine intelligence isn't happening in some distant future, it's the reality millions of patients encounter today, often without realizing an algorithm has influenced their care. The ethical frameworks, governance structures, and accountability mechanisms explored in previous chapters now meet the lived experience of real people facing real health decisions, where the stakes couldn't be higher.

Artificial intelligence has moved from research laboratories into examination rooms, operating theaters, pharmaceutical development facilities, and the devices we wear on our wrists. Machine learning algorithms now detect cancers in mammograms, predict which patients will develop sepsis, recommend treatment protocols for complex conditions, accelerate drug discovery timelines from decades to months, and promise interventions that might extend human health span in ways previous generations could barely imagine. These applications represent some of AI's most compelling benefits: technologies that could save lives, reduce suffering, and democratize access to expert-level medical knowledge.

Yet these same technologies raise urgent ethical questions that demand our attention now, before patterns become entrenched and inequities become irreversible. When an algorithm recommends one treatment over another, who bears responsibility if the outcome proves harmful? When AI-driven diagnostic tools perform brilliantly on some populations but fail others, who ensures equitable access and accuracy? When longevity interventions developed through AI-accelerated research become available, will they extend healthy life for everyone or only for those who can afford them? When our smartphones and wearables collect intimate health data to power AI predictions, who protects our privacy and prevents discrimination based on algorithmic assessments of our bodies?

These questions extend beyond hospital walls into the fabric of our daily lives. AI doesn't just influence medicine in clinical settings; it shapes how we understand our own health through fitness trackers that analyze our sleep patterns, apps that assess our mental well-being, smart home devices that monitor our movements, and algorithms that determine which health information we see online. The boundary between medical AI and everyday AI has blurred, creating an environment where artificial intelligence continuously observes, analyzes, and influences our health-related choices, often invisibly.

The ethical principles examined throughout this book, fairness, accountability, transparency, privacy, and human dignity, take on visceral urgency when applied to healthcare and daily life. A biased hiring algorithm causes injustice, but a biased diagnostic algorithm can cost lives. A privacy breach in social media feels invasive, but unauthorized access to health data can destroy insurability, employability, and fundamental security. The governance gaps that allow problematic AI in other sectors become matters of life and death when applied to medicine.

This chapter examines how we navigate these challenges practically, moving from abstract ethical principles to concrete decisions that patients, healthcare professionals, technology users, and policymakers face today. It explores both the remarkable potential of AI in medicine and longevity science and the ethical guardrails necessary to ensure these technologies serve human flourishing rather than corporate profit or algorithmic efficiency. Drawing on real-world implementations, documented failures, and emerging best practices, we'll investigate how to harness AI's diagnostic power while preserving clinical judgment, how to accelerate medical innovation while ensuring equitable access, and how to benefit from health-monitoring technologies while protecting privacy and autonomy.

The choices we make now about AI in healthcare and everyday life will determine whether these technologies reduce health disparities or deepen them, whether they enhance human connection or replace it, and whether they extend dignified life for all or only for the privileged few. Understanding how to recognize ethical AI applications and demand better when systems fall short isn't just theoretical knowledge; it's practical wisdom we need to protect ourselves, our families, and our communities in an AI-saturated world.

AI in Medical Diagnosis: Accuracy, Accountability, and the Human Element

Artificial intelligence has achieved remarkable accuracy in medical diagnosis, often surpassing human clinicians in specific detection tasks. In a large-scale study involving over **52,000** patients, an AI system reached **85.7%** accuracy in predicting medical outcomes, with **86.3%** sensitivity and **85.7%** specificity, significantly outperforming trained nephrologists who achieved **69.4%** accuracy, **72%** sensitivity, and 66%

specificity. 97 For colon cancer detection, AI algorithms have demonstrated 98% accuracy compared to 96.9% for trained pathologists, while AI screening for retinal disorders and lung cancer has reached 95.2% and 98.7% accuracy, respectively. 98, 99 These numbers represent genuine breakthroughs that could save lives through earlier detection and more precise diagnosis.

Yet these impressive statistics mask a more complex reality about how AI functions in clinical settings and who bears responsibility when algorithms make mistakes. The accuracy of AI systems depends heavily on the quality and representativeness of their training data, and performance can vary dramatically across different patient populations. When researchers tested AI diagnostic tools across demographic groups, they found that Black respondents selected AI diagnostic tools less often than White respondents, suggesting trust disparities rooted in historical medical injustices. 98 More troubling, some AI models showed improved precision for certain racial groups while declining for others, revealing how algorithmic bias can create unequal diagnostic accuracy based on patient demographics. 98

The question of accountability becomes particularly urgent when AI makes errors with serious consequences. In a multicenter randomized clinical vignette survey, researchers found that when AI predictions were systematically biased, clinician accuracy dropped by 11.3 percentage points, and providing explanations for the AI's reasoning did not mitigate this negative effect. 96 This finding challenges the common assumption that making AI more transparent automatically makes it safer. When clinicians receive flawed recommendations from systems they've been trained to trust, even with explanations, they may follow those recommendations to patients' detriment. Who bears responsibility in such cases, the algorithm developers, the healthcare institution that deployed it, or the clinician who relied on its output?

The most effective diagnostic outcomes emerge not from AI replacing human judgment but from thoughtful collaboration between algorithmic analysis and clinical expertise. When clinicians in one study were provided with AI predictions accompanied by clear explanations, their diagnostic accuracy increased by 4.4 percentage points compared to baseline performance. 96 This improvement demonstrates AI's potential as what researchers call a "reliable assistant," a tool that flags abnormalities and provides data-driven insights while human clinicians apply contextual judgment, consider patient history, and make final decisions. 99

This collaborative model preserves what Dr. Fei-Fei Li, co-director of Stanford's Human-Centered AI Institute, describes as essential: "AI in medicine must be designed with the human in the loop, not as a replacement for human judgment." 100 Radiologists reviewing AI-flagged scans bring years of training that allows them to distinguish between genuinely concerning findings and artifacts the algorithm misinterprets. Physicians considering AI-generated treatment recommendations understand patient preferences, comorbidities, and social circumstances that no algorithm can fully capture. The human element provides not just clinical expertise but also empathy, communication, and ethical reasoning that remain irreplaceable in healthcare.

The challenge facing healthcare systems today involves creating structures that harness AI's analytical power while maintaining clear accountability and preserving human oversight. This requires training clinicians to critically evaluate AI outputs rather than accept them uncritically, implementing monitoring systems that detect algorithmic bias across patient populations, and establishing legal and ethical frameworks that clearly assign responsibility when AI-assisted decisions cause harm. Technology's accuracy matters less than how thoughtfully it's integrated into care delivery and who ensures it serves all patients equitably.

Algorithmic Treatment Recommendations: Who Decides Your Care Plan?

When a physician opens a patient's electronic health record today, the treatment recommendations that appear on screen increasingly originate not from clinical guidelines alone but from algorithms analyzing thousands of data points, genetic markers, lab results, medication histories, demographic information, and patterns extracted from millions of similar cases. These AI-driven clinical decision support systems now influence care plans for conditions ranging from cancer treatment selection to antibiotic prescribing, fundamentally altering the question of who decides what care a patient receives.

The promise of algorithmic treatment recommendations centers on personalization and precision. The Clinical Pharmacogenetics Implementation Consortium provides guidelines for drug prescriptions tailored to patients' genetic profiles, improving both safety and effectiveness in medications like warfarin, where dosing varies dramatically based on genetic variants. **101 In** pediatric cancer care, genomic analysis powered by AI has enabled targeted therapy for specific subgroups of medulloblastoma patients, reducing the need for more harmful treatments while improving outcomes. **101** These applications demonstrate AI's capacity to integrate complex biological data in ways that exceed human cognitive processing, potentially delivering "the right treatment, at the right time, for the right patient". **101, 102**

Yet the integration of algorithms into treatment decisions creates profound questions about authority, accountability, and patient autonomy. Most current systems function as clinical decision support tools—designed to augment rather than replace physician judgment. 101 102 105 The ideal represents a hybrid model where AI provides evidence-based recommendations derived from vast datasets while clinicians interpret these suggestions within the context of patient preferences, values, and unique circumstances. **101** This partnership preserves human oversight while leveraging computational power.

The reality, however, often proves more complicated. Research has documented how algorithmic recommendations can subtly shift decision-making authority away from both clinicians and patients. When a widely used algorithm for predicting healthcare needs in the United States was examined in 2019, researchers discovered it systematically underestimated care requirements for Black patients compared to equally sick white patients. **12, 106** The algorithm had learned to use healthcare spending as a proxy for health needs, faithfully replicating a system that historically spent less on Black patients. Hospitals relying on this algorithm's recommendations inadvertently perpetuated inequitable care, demonstrating how algorithmic bias can determine who receives interventions and who gets overlooked.

The transparency challenge compounds these equity concerns. Many AI systems function as "black boxes," making it difficult for clinicians to understand how recommendations are generated or to explain them to patients. **12, 106** When a physician cannot articulate why an algorithm suggests one chemotherapy regimen over another, the foundation for informed consent and shared decision-making erodes. Patients may find themselves following treatment plans shaped by opaque computational processes they never consented to and their doctors cannot fully explain.

Accountability becomes particularly murky when algorithmic recommendations lead to poor outcomes. Determining responsibility between the algorithm developers, the healthcare institution that deployed it, the clinician who followed its guidance, and the patient who consented to treatment presents legal and ethical challenges that existing frameworks struggle to address. Regulatory agencies like the FDA are developing oversight mechanisms for AI-enabled medical devices **103, 104**, but the pace of algorithmic development outstrips regulatory capacity, creating gaps in evaluation and monitoring. **103, 104, 107**

Insurance companies further complicate the decision-making landscape by increasingly using algorithms to determine coverage and approve or deny treatments. 106 These systems can override clinical judgment by refusing to pay for recommended interventions, effectively deciding care plans through financial gatekeeping rather than medical reasoning. The criteria these algorithms use often remain proprietary, leaving both patients and physicians unable to challenge denials meaningfully.

The question of who decides a patient's care plan at an algorithmic age demands urgent attention. Ensuring that AI augments rather than supplants human judgment, that algorithms serve all patients equitably, and that decision-making authority remains transparent and accountable requires deliberate design choices, robust oversight, and sustained vigilance from everyone involved in healthcare delivery.

AI-Driven Drug Discovery and Longevity Research: Extending Life, But for Whom?

Artificial intelligence is compressing what once took a decade and billions of dollars into months, fundamentally transforming how researchers discover new drugs and approach aging itself. Traditional pharmaceutical development demanded approximately ten years of testing and enormous financial investment before a single drug reached patients. 111 AI changes this calculus by enabling researchers to analyze vast databases of compounds and proteins, identifying promising therapeutic targets with unprecedented efficiency and speed. 111 This acceleration represents genuine scientific progress—but it also raises an urgent question about who will ultimately benefit from these life-extending innovations.

The shift toward AI-driven drug discovery has enabled a conceptual revolution in how researchers approach aging. Rather than the traditional "one-drug, one-target" paradigm, scientists now pursue polypharmacological targeting, recognizing that aging affects multiple biological systems simultaneously and may require interventions that work across several pathways. 108 In 2025, researchers at Scripps Research and Gero demonstrated this principle using AI to identify drugs targeting dopamine, serotonin, and histamine receptors, all linked to aging processes. The results proved remarkable: more than seventy percent of the anti-aging drugs identified by the AI tool significantly extended the lifespan of laboratory worms, with one novel compound increasing lifespan by seventy-four percent. **108**

Michael Petrascheck, professor at Scripps Research and co-senior author of the study, explained the significance: "This study shows that artificial intelligence can help us go beyond the traditional 'one-drug, one-target' mindset. By embracing the complexity of polypharmacological targeting, we were able to identify compounds that produce stronger and more reliable effects on lifespan than anything we've seen in previous screens." **108** This represents not merely faster discovery within existing approaches but a qualitative shift in what researchers can identify and test.

Companies like Insilico Medicine have structured entire operations around AI-driven longevity research, with thirty AI-based projects specifically chosen for their "dual purpose" potential, targeting both disease and aging simultaneously. **109** The company's PandaOmics platform integrates more than twenty advanced AI models to identify therapeutic targets, discovering **145** potential dual-purpose aging targets by **2022.** 110 In silico achieved first-in-class proof of concept with Renteria, which demonstrated dose-dependent efficacy against idiopathic pulmonary fibrosis in a Phase Idia clinical trial while targeting a leading age-associated gene linked to multiple hallmarks of aging. **110**

Beyond drug discovery, AI enables early detection of disease patterns through analysis of medical records, wearable device data, and genomic information, spotting conditions like cancer or cardiovascular disease based on complex biomarkers that human clinicians might miss. 111 Deep learning systems are revolutionizing biomarker discovery and developing "aging clocks," mathematical models that measure biological age more accurately than chronological age, enabling researchers to assess whether interventions genuinely slow aging at the molecular level. 113

Yet the acceleration of discovery does not automatically resolve questions of access and equity. Historical patterns in pharmaceutical distribution suggest that breakthrough therapies often reach wealthy populations first, if they reach broader populations at all. Alex Zahvoronkov, founder and CEO of Insilico Medicine, acknowledges a crucial threshold that has not yet been crossed: "The first AI-discovered drug that gets approved to fight a disease and subsequently is able to demonstrate credible reversal of aging biomarkers in a real clinical trial will be a major milestone." 109 This milestone matters not merely for scientific validation but because it will establish precedent for how regulatory bodies, insurers, and healthcare systems evaluate and distribute AI-discovered anti-aging therapeutics.

The impact of AI-driven longevity interventions will likely be uneven across different populations, and the magnitude and timing of these changes remain uncertain. 111 A critical distinction has emerged between extending lifespan and compressing morbidity, delaying the onset of chronic, age-related diseases so that people remain healthy longer. 112 A therapy that extends life while prolonging disease and suffering differs fundamentally from one that delays disease onset, allowing individuals to maintain quality of life. The choices made now about clinical trial design, pricing, distribution, and access will determine whether AI-driven longevity research becomes a tool for universal health improvement or another mechanism through which technological benefits accrue primarily to the already privileged.

Health Equity and Access: Ensuring AI Medicine Benefits Everyone, Not Just the Privileged

The acceleration of AI-driven medical breakthroughs documented in drug discovery and longevity research confronts an uncomfortable reality: technological progress does not automatically translate into equitable access. History demonstrates repeatedly that medical innovations often reach wealthy populations first, if they reach broader populations at all. The question facing healthcare systems today is whether AI medicine will narrow health disparities or widen them, whether these powerful tools will serve everyone or primarily those who can afford them.

The challenge begins with the data that trains medical AI systems. When algorithms learn from datasets that underrepresent certain populations, they perform poorly for those groups in clinical application. A 2019 study published in science examined a widely used algorithm for predicting which patients needed additional medical care and discovered systematically under-prioritized Black patients compared to equally sick white patients. 4 The algorithm had learned to use healthcare spending as a proxy for health needs, faithfully replicating a system that historically spent less on Black patients. This bias affected millions of patients, demonstrating how AI can automate and scale existing inequalities unless developers deliberately ensure training data represents diverse populations.

The digital divide creates another barrier to equitable access. AI-powered diagnostic tools, telemedicine platforms, wearable health monitors, and smartphone applications require infrastructure that remains unevenly distributed. Rural communities often lack the broadband connectivity necessary for AI-enhanced telemedicine, while low-income populations may not own the devices required to access AI health applications. When healthcare systems deploy AI tools without addressing these access barriers, they risk creating a two-tiered system where privileged patients receive cutting-edge AI-assisted care while others continue with traditional approaches; or receive no care at all.

The commercialization of AI health technologies compounds these equity concerns. Private companies developing AI medical tools face pressure to generate returns for investors, creating incentives to focus on markets with higher purchasing power. Longevity interventions discovered through AI-accelerated research may be priced beyond the reach of most people, available only to wealthy individuals who can afford experimental treatments. Dr. Fei-Fei Li, Co-Director of Stanford's Human-Centered AI Institute, emphasizes this challenge: "AI in healthcare must be designed with equity at its core. If we don't, we risk building a future where only the privileged benefit from the latest medical breakthroughs." **2**

Addressing these disparities requires deliberate intervention at multiple levels. Inclusive data collection represents a foundational step, ensuring training datasets reflect the demographic diversity of populations these systems will serve. The National Institutes of Health's All of Us Research Program exemplifies this approach, gathering health data from diverse participants specifically to support equitable AI research. **3** Community engagement proves equally critical, involving patients and community leaders in designing and deploying AI health tools rather than imposing solutions developed without input from affected populations.

Regulatory frameworks must mandate equity impact assessments for AI health technologies, requiring developers to demonstrate their systems perform equitably across demographic groups before deployment. Dr. Ruha Benjamin, Professor of African American Studies at Princeton University, argues, "We need regulatory frameworks that require transparency and accountability in AI-driven healthcare, especially regarding equity and access." **67** Such oversight can prevent biased systems from reaching clinical practice and ensure ongoing monitoring detects disparities that emerge over time.

Practical deployment strategies matter as much as technical development. Subsidizing or providing free access to AI health tools for underserved populations, developing low-bandwidth versions for areas with limited connectivity, and investing in public education about AI in healthcare can help bridge access gaps. The deployment of AI-powered diabetic retinopathy screening in rural Indian clinics demonstrates this potential, automating diagnosis to reach thousands of patients who lack access to ophthalmologists, though challenges remain in ensuring cultural appropriateness and health literacy.

The choices made now about data collection, algorithm design, regulatory oversight, pricing, and distribution will determine whether AI medicine becomes a tool for reducing health disparities or another mechanism through which technological benefits accrue primarily to those already privileged. Health equity in the age of AI is not an inevitable outcome but a goal that requires sustained commitment, deliberate design, and vigilant oversight from everyone involved in developing and deploying these powerful technologies.

AI in Everyday Health Decisions: Wearables, Apps, and the Quantified Self

The smartphone in your pocket, the watch on your wrist, and the app tracking your sleep patterns represent a profound shift in how health decisions happen, not in doctor's offices during scheduled appointments, but continuously, algorithmically, and often invisibly throughout daily life. Artificial intelligence embedded in consumer devices now monitors heart rhythms, analyzes sleep quality, predicts fertility windows, assesses mental health, and recommends behavioral changes based on patterns extracted from millions of users. This transformation of everyday health management through wearables, apps, and quantified self-movement promises unprecedented personal empowerment while raising urgent questions about privacy, autonomy, and who truly benefits from the intimate data these systems collect. 4, 2, 3

The quantified self-movement, the practice of using digital technologies to collect, analyze, and act upon personal health data, has moved from niche enthusiasm to mainstream behavior. Wearable devices continuously gather biometric information, including heart rate, movement patterns, sleep stages, blood oxygen levels, and increasingly sophisticated metrics like heart rate variability and electrodermal activity. Mobile health applications apply machine learning algorithms to detect patterns, flag anomalies, and deliver personalized recommendations. The Apple Watch's ability to detect atrial fibrillation demonstrates this potential: AI algorithms analyzing heart rhythm data in real-time have alerted users to irregular heartbeats, leading some to seek medical evaluation and receive potentially life-saving treatment for previously undiagnosed cardiac conditions.

This continuous health monitoring enables early detection of problems that might otherwise go unnoticed until symptoms become severe. AI-powered apps can identify sleep apnea patterns, detect changes in gait that might indicate neurological issues, or recognize behavioral shifts associated with depression. The personalization these systems offer, tailoring advice to individual patterns rather than population averages, represents genuine progress toward preventive, data-driven healthcare that extends beyond clinical settings into the fabric of daily existence.

Yet the ethical challenges embedded in these everyday health technologies demand scrutiny equal to their clinical counterparts. Privacy concerns prove particularly acute because wearables and apps collect some of the most intimate data imaginable: information about bodies, behaviors, locations, and patterns that reveal far more than users typically realize. 2, 3 When fitness app Strava published a global heatmap of user activity in 2018, researchers discovered it inadvertently revealed locations of military bases and personnel, demonstrating how aggregated health and fitness data can expose sensitive information never intended for public view. Many devices transmit data to cloud servers, where it becomes vulnerable to breaches, unauthorized access, and uses that extend far beyond the health insights users sought.

The question of informed consent becomes murky when users click "agree" on lengthy terms of service without understanding what data is collected, how algorithms analyze it, who gains access, or how it might be used for purposes beyond personal health tracking. Insurance companies have begun offering premium discounts for customers who share fitness tracker data, creating financial incentives that blur the line between voluntary health monitoring and coercive surveillance. Employers increasingly provide wearables to employees, framing them as wellness benefits while gathering data that could inform hiring, promotion, or termination decisions.

Algorithmic bias affects everyday health AI just as it does clinical systems. When wearables and apps are trained on non-representative datasets, they perform less accurately for certain demographic groups. [2] Research has documented that some heart rate monitors and pulse oximeters, technologies that underpin many wearable health features, show reduced accuracy on darker skin tones, and have a bias rooted in development and testing processes that failed to include diverse populations. These disparities mean the quantified self-movement may provide reliable insights for some users while delivering flawed data to others, potentially exacerbating rather than reducing health inequities.

The autonomy implications of continuous algorithmic feedback deserve careful consideration. When devices constantly nudge users toward behavior changes, more steps, better sleep, and different eating patterns, the line between helpful guidance and psychological manipulation becomes difficult to discern. The gamification strategies many apps employ, designed to increase engagement through points, streaks, and social comparison, can transform health management from autonomous decision-making into compulsive response to algorithmic prompts. For some users, particularly those vulnerable to obsessive behaviors, the quantified self can become a source of anxiety rather than empowerment, with AI-driven feedback loops reinforcing unhealthy fixations on metrics.

Accountability remains unclear when AI-powered health apps provide recommendations that prove harmful. Unlike regulated medical devices subject to FDA oversight, many consumer health apps operate in a regulatory gray zone where responsibility for accuracy, safety, and efficacy remains ambiguous. When an app's algorithm suggests a user's symptoms don't warrant medical attention and that assessment proves dangerously wrong, determining whether the developer, the user, or the algorithm itself bears responsibility presents legal and ethical challenges that existing frameworks struggle to address.

The transparency problem compounds these concerns. **3** Most health AI operates as a black box, making it impossible for users to understand how recommendations are generated or to evaluate their reliability. When a sleep app suggests changes to bedtime routines or a mental health app assesses emotional state, users typically cannot access the logic behind these determinations, the data used to train the models, or the accuracy rates across different populations. This opacity undermines informed decision-making and prevents users from critically evaluating the guidance they receive.

The everyday health AI embedded in wearables and apps represents neither pure benefit nor pure threat but rather a complex technology whose impact depends entirely on how thoughtfully it's designed, deployed, and regulated. Ensuring these tools serve human flourishing rather than corporate data collection requires robust privacy protection **2**, transparent algorithms, inclusive development processes that address bias, clear accountability structures, and genuine informed consent that respects user autonomy. The choices individuals make about adopting these technologies and the standards society demands from companies developing them will determine whether the quantified self-enhances health equity and personal agency or becomes another mechanism through which surveillance and algorithmic control extend into the most intimate aspects of human existence. **4, 2, 3, 114, 115**

Preserving Human Connection in AI-Enhanced Healthcare: When Algorithms Can't Replace Compassion

The most sophisticated diagnostic algorithm cannot hold a frightened patient's hand, cannot recognize the unspoken worry behind a routine question, and cannot adjust its tone when delivering difficult news based on subtle cues of emotional readiness. As artificial intelligence assumes greater roles in healthcare delivery, from triaging patient messages to recommending treatment protocols, the irreplaceable value of human connection in healing becomes both more apparent and more vulnerable.

Research consistently demonstrates that compassion and empathy constitute essential components of effective medical care, not merely pleasant additions to technical competence **119 and 120**. The healing relationship between patient and clinician significantly affects treatment outcomes, patient satisfaction, and adherence to medical recommendations **119**. Yet these uniquely human capacities remain beyond the reach of current AI technologies, which can process data and generate responses but lack authentic emotional understanding and the moral reasoning that underpins genuine care **119, 120**

Healthcare institutions increasingly deploy AI to handle routine patient communications through portal messages, chatbots, and automated responses. A **2025** study examining patient reactions to AI-generated responses found that while patients appreciated detailed and empathetic replies to straightforward health queries, they emphasized the importance of transparency, wanting to know when AI rather than a human clinician was responding, particularly for sensitive or complex situations. **116** The research revealed a crucial distinction: patients may accept AI assistance for administrative or routine matters, but they expect and value human involvement for communications that require trust, nuance, and the intimacy inherent in the patient-clinician relationship. **116**

The American Nurses Association has articulated clear ethical boundaries for AI integration in nursing practice, emphasizing that artificial intelligence should augment and support clinical care without compromising the human interactions central to the nurse-patient relationship. **120** This principle extends across all healthcare professions; AI should function as a tool that enhances rather than replaces the core relational aspects of medicine. **120** When algorithms handle repetitive administrative tasks, they can reduce clinician burnout and create more time for direct patient engagement **116, 121**, but when they substitute for human presence in moments requiring compassion, they risk eroding the foundation of therapeutic relationships. **119 120 121**

The World Health Organization has established six key principles for ethical AI in healthcare: protecting human autonomy, promoting human well-being and safety, ensuring transparency and explainability, fostering responsibility and accountability, ensuring inclusiveness and equity, and promoting AI that is responsive and sustainable **117**. These principles recognize that technology must serve human values rather than compromise them and that the benefits of efficiency cannot justify the loss of elements essential to healing.

At the University of California, San Francisco, the implementation of AI-powered medical scribes demonstrates a thoughtful approach to preserving human connection while leveraging technological capabilities. **118** The institution established governance committees bringing together clinicians, ethicists, data scientists, and patient advocates to evaluate AI tools before deployment, ensuring they enhance rather than diminish the quality of patient care. [118] This model allows physicians to maintain eye contact and engagement with patients during consultations while AI handles documentation, using technology to support the relationship rather than intruding upon it. [118]

The challenge facing healthcare systems involves drawing clear boundaries around AI's appropriate role. Algorithms excel at processing information, identifying patterns in vast datasets, and handling routine communications that consume clinician time without requiring human judgment or emotional intelligence. [116, 121] They fail, however, at recognizing the unspoken needs of a patient struggling to articulate fears, at adapting communication style based on cultural context and individual personality, at providing the physical presence and touch that convey care beyond words. [120]

Preserving human connection in AI-enhanced healthcare requires deliberate choices about where algorithms assist and where they must not intrude. It demands transparency so patients know when they interact with machines rather than humans. [116, 117, 122] It necessitates protecting time and space for the healing of relationships that remain medicine's most powerful and irreplaceable tool—the authentic human presence that no algorithm can replicate. [119, 120, 121] The transformation of healthcare through artificial intelligence represents one of the most consequential applications of algorithmic systems in human society. From diagnostic algorithms that detect cancers invisible to human eyes to AI-accelerated drug discovery that compresses decades of research into months, these technologies demonstrate genuine potential to reduce suffering, extend healthy life, and democratize access to expert medical knowledge. The wearables on our wrists, the apps on our phones, and the algorithms embedded in electronic health records have moved AI from abstract concern to daily reality, influencing health decisions for millions of people who may not even realize an algorithm has shaped their care.

Yet this chapter has revealed that technological capability alone determines nothing about outcomes. The same AI systems that achieve remarkable diagnostic accuracy can perpetuate deadly biases when trained on non-representative data, systematically under-serving populations already marginalized by healthcare inequities. The algorithms that promise personalized treatment recommendations can erode patient autonomy and clinical judgment when deployed without transparency or accountability. The longevity interventions discovered through AI-driven research may extend healthy life for everyone or only for those privileged enough to afford them, depending entirely on choices made now about access, pricing, and distribution. The quantified self-movement enabled by consumer health technologies offers unprecedented personal health insights while simultaneously creating new vulnerabilities to surveillance, manipulation, and discrimination based on intimate biometric data.

The ethical frameworks explored throughout this book, fairness, accountability, transparency, privacy, and human dignity, take on visceral urgency when applied to medicine and everyday health. Several principles emerge as essential for ensuring AI serves human flourishing rather than corporate profit or algorithmic efficiency:

1. AI must function as a collaborative tool that augments human expertise rather than replaces clinical judgment, preserving the irreplaceable elements of compassion, contextual understanding, and moral reasoning that remain central to healing
2. Training data must represent the full diversity of populations these systems will serve, with ongoing monitoring to detect and correct biases that create unequal accuracy or outcomes across demographic groups
3. Transparency and explainability are not optional features but fundamental requirements, enabling both clinicians and patients to understand how recommendations are generated and to make truly informed decisions about care

4. Privacy protections must extend beyond preventing data breaches to addressing how health information collected by wearables and apps can be used for insurance discrimination, employment decisions, and other purposes that threaten human dignity
5. Regulatory frameworks must mandate equity impact assessments and establish clear accountability when AI systems cause harm, closing the gaps that currently allow problematic algorithms to reach clinical practice without adequate oversight

The healthcare professionals, patients, technology developers, and policymakers navigating this landscape face decisions that will reverberate for generations. Physicians must learn to critically evaluate algorithmic recommendations rather than accept them uncritically, preserving their role as advocates for individual patients against systems optimized for population-level efficiency. Patients must demand transparency about when and how AI influences their care, insisting on genuine informed consent rather than passive acceptance of opaque algorithmic decisions. Developers must prioritize equity and safety over speed to market, recognizing that biased or flawed medical AI can cost lives. Regulators must close the governance gaps that allow consumer health technologies to operate without the oversight applied to medical devices, establishing standards that protect users without stifling beneficial innovation.

The promise of AI in medicine and everyday health remains genuine; these technologies can detect diseases earlier, personalize treatments more precisely, accelerate scientific discovery, and extend the reach of expert care to underserved populations. Realizing this promise while avoiding the very real dangers of bias, inequity, privacy erosion, and dehumanization requires vigilance, deliberate design choices, and sustained commitment to ensuring technology serves human values rather than compromises them. The algorithms analyzing our health data, recommending our treatments, and monitoring our bodies will reflect the priorities we embed in them through the choices we make today.

CHAPTER 7: CHOOSING THE WORLD WE WANT: A CALL TO ETHICAL ACTION

Standing in my garden on our Kentucky farm, watching new growth emerge from soil we've carefully tended, I'm reminded that the future, like a garden, doesn't simply happen to us. After three decades as a trauma nurse, I learned that the difference between tragedy and recovery often comes down to the actions people take in critical moments, and we are living through such a moment now as artificial intelligence reshapes the foundations of human society. The choices we make in the next decade about artificial intelligence will echo for generations, determining whether this technology becomes humanity's greatest tool for flourishing or a force that diminishes what makes us human. This is not hyperbole; it is the sober assessment of researchers, ethicists, and technologists who understand that we are at an inflection point where the systems being designed today will shape the social, economic, and political landscape of tomorrow.

Throughout this book, we have examined how AI reflects human values, how bias becomes embedded in algorithms, what responsible development requires, how privacy and dignity are threatened, who governs these powerful systems, and how AI is transforming medicine and daily life. Each chapter has revealed both extraordinary potential and serious risks. The question that remains is perhaps the most important one: What will you do with this knowledge?

The temptation to feel powerless in the face of massive technology companies, complex algorithms, and rapid innovation is understandable. Many people believe that AI's trajectory is predetermined, that decisions about these systems happen in boardrooms and research labs far removed from ordinary life, and that individual voices cannot possibly matter against such powerful forces. This belief is not only wrong. It is dangerous. History demonstrates repeatedly that technological systems reflect the values and priorities of those who show up to shape them and that the absence of diverse voices leads to technologies that serve narrow interests rather than the common good.

The window for meaningful influence is open, but it will not remain open indefinitely. AI systems are being deployed at unprecedented speed across healthcare, education, criminal justice, employment, finance, and virtually every domain of human activity. Each deployment creates precedents, establishes norms, generates data that trains future systems, and builds infrastructure that becomes progressively harder to change. The decisions being made now about privacy protections, bias mitigation, transparency requirements, and accountability mechanisms will be far easier to implement today than to retrofit into entrenched systems tomorrow.

This urgency does not require panic; it requires purposeful action informed by clear ethical principles. The frameworks explored in previous chapters provide tools for recognizing when AI is being used without ethics, compassion, or responsibility. The case studies have illustrated what goes wrong when systems are deployed without adequate consideration of their impact on human dignity, equity, and autonomy. The question is whether we will apply these lessons before harm becomes widespread and irreversible, or whether we will look back with regret at opportunities missed.

The good news is that pathways for influence exist at every level. As individuals, we make daily choices about which technologies to use, which companies to support, and how to protect our privacy and autonomy. As professionals, we can advocate for ethical practices within our organizations, refuse to implement systems that violate our values, and demand accountability from leadership. As citizens, we can engage with policymakers, support regulations that protect human dignity, and participate in public discourse about AI governance. As community members, we can educate others, build coalitions around shared values, and create local initiatives that demonstrate what ethical AI looks like in practice.

None of these actions alone will determine AI's future, but collectively they create the conditions for technology that serves humanity rather than diminishes it. The world we want, one where AI enhances human capability without eroding human dignity, where innovation proceeds with wisdom rather than reckless speed, and where technology's benefits are shared equitably rather than concentrated among the privileged, will not emerge by accident. It requires conscious choice, sustained effort, and the courage to act even when outcomes are uncertain.

This chapter provides practical guidance for becoming an active participant in shaping AI's trajectory, exploring concrete actions across different spheres of influence and offering frameworks for ethical decision-making in an AI-saturated world.

Individual Agency in the AI Age: How Your Choices Shape Technology's Future

The belief that individual actions cannot influence the trajectory of artificial intelligence represents one of the most dangerous misconceptions of our time. Research from the Oxford Internet Institute demonstrates that technology adoption patterns, user resistance, and consumer choices have historically shaped which innovations succeed and which fail, even when backed by powerful corporations. The future of AI is not predetermined by tech giants or government agencies alone; it is being written through millions of daily decisions made by individuals who use, reject, question, or demand changes to these systems. 4, 2

Every interaction with AI-powered technology constitutes a choice with ethical implications. When someone accepts default privacy settings without review, shares personal data in exchange for convenience, or uses facial recognition to unlock a phone, they participate in normalizing certain technological practices while making others economically viable 4, 2. Conversely, when users delete apps over privacy concerns, demand explanations for algorithmic decisions, or choose services based on ethical considerations, they send market signals that influence corporate behavior 4, 2. The **2018** Google employee walkout, which drew over **20,000** participants protesting the company's handling of sexual harassment and involvement in military AI projects, demonstrated how collective individual action can force policy changes even within the world's most powerful technology companies **67**.

Consumer choices extend beyond simple adoption or rejection. The questions individuals ask about AI systems, the complaints they file, and the alternatives they seek create pressure for transparency and accountability **2, 67**. When San Francisco and Boston banned government use of facial recognition technology, these policy victories emerged from sustained civic engagement by ordinary citizens who attended public meetings, contacted representatives, and organized community opposition **67**. Research from MIT's Media Lab shows that public resistance to biased or invasive AI applications has directly influenced corporate decisions to delay or modify product launches, proving that individual voices aggregated through civic channels can alter technological development.

Professional agencies carry weight in shaping AI's future. Developers, data scientists, and engineers make countless micro-decisions during system design that embed values into technology, such as which datasets to use, which biases to test for, which explanations to provide, and which safeguards to implement **2, 67**. The ACM Code of Ethics emphasizes that computing professionals bear responsibility for the foreseeable consequences of their work, including the obligation to refuse participation in projects that violate ethical principles 67. When AI researchers at major universities declined military funding for autonomous weapons research, their individual choices influenced institutional policies and contributed to broader debates about AI's role in warfare 2, 67.

The European Union's General Data Protection Regulation, which includes provisions requiring explanations for automated decisions affecting individuals, emerged partly from sustained citizen advocacy for privacy rights 67. This "right to explanation" represents a concrete example of how individual demands for agency over personal data can reshape the legal landscape governing AI systems 67. Studies from the Berkman Klein Center at Harvard demonstrate that jurisdictions with active citizen engagement in technology policy tend to develop more robust protections for individual rights in AI contexts 4, 67.

Individual agency also manifests in educational choices and literacy development 67. People who invest time in understanding how AI systems work, what data they collect, and what rights they possess become better equipped to make informed decisions and advocate effectively 67. Research published in Nature Machine Intelligence indicates that even basic AI literacy significantly increases individuals' capacity to identify problematic applications, demand accountability, and participate meaningfully in governance discussions 67.

The cumulative effect of individual choices creates the social and economic conditions that either enable or constrain AI development 4, 67. Technology companies respond to market signals, policymakers respond to constituent pressure, and developers respond to professional norms, all of which are shaped by the aggregated decisions of individuals who refuse to accept that their choices do not matter 2, 67. The world we want will not emerge from passivity but from the conscious exercise of agency by people who understand that their decisions, multiplied across millions of others, possess the power to shape technology's future 4, 67.

Organizational Leadership and Ethical AI Implementation: Creating Change from Within

While individual choices matter, the organizations where people work represent concentrated points of leverage for shaping AI's ethical trajectory. Leaders within corporations, healthcare systems, educational institutions, and government agencies possess unique power to influence how AI is developed, deployed, and governed at scale. Research from MIT's Sloan School of Management demonstrates that organizational culture and leadership commitment are stronger predictors of ethical AI implementation than technical capabilities or regulatory compliance alone. The question for professionals at every level is not whether they can influence their organization's approach to AI, but whether they will choose to exercise that influence.

Organizational leadership in ethical AI extends far beyond the executive suite. While C-suite commitment provides essential resources and legitimacy, meaningful change requires champions throughout the organizational hierarchy, middle managers who shape team priorities, technical leads who make design decisions, procurement officers who select vendors, and frontline employees who implement systems and witness their impacts. The IEEE Global Initiative on Ethics of Autonomous and Intelligent Systems emphasizes that ethical AI implementation succeeds when responsibility is distributed across roles rather than concentrated in a single ethics officer or committee. Organizations that treat ethics as everyone's responsibility rather than a specialized function demonstrate better outcomes in bias mitigation, transparency, and accountability.

Creating change from within begins with establishing clear ethical principles that guide AI decisions. Microsoft's AI Principles, developed following internal debate and external scrutiny, provide a concrete example of how organizations can articulate commitments to fairness, reliability, privacy, inclusiveness, transparency, and accountability. These principles function not as abstract aspirations but as decision-making frameworks that team's reference when evaluating whether to pursue specific AI applications. Research published in the Harvard Business Review indicates that organizations with explicit AI ethics principles demonstrate measurably different development practices compared to those relying on implicit norms or compliance-driven approaches.

Governance structures translate principles into practice. Effective AI governance requires cross-functional teams that bring together technical expertise, domain knowledge, legal understanding, and ethical reasoning. **IBM's AI Ethics** Board exemplifies this multidisciplinary approach, reviewing high-impact AI projects and providing recommendations before deployment. The board's composition includes representatives from engineering, research, legal, policy, and business units, ensuring that ethical considerations are integrated with technical feasibility and business objectives. Studies from the Berkman Klein Center at Harvard demonstrate that organizations with formal ethics review processes identify and address potential harm earlier in development cycles, reducing costly redesigns and reputational damage.

Transparency and accountability mechanisms ensure that ethical commitments translate into measurable outcomes. Regular algorithmic audits, bias testing, and impact assessments provide evidence of whether AI systems perform as intended across diverse populations. Google's Model Cards and similar documentation frameworks require teams to disclose training data characteristics, performance metrics across demographic groups, and known limitations. This transparency serves multiple purposes: it enables internal accountability, builds external trust, and creates records that facilitate learning when systems fail. Research from Stanford's Institute for Human-Centered Artificial

Intelligence shows that organizations practicing systematic transparency demonstrate better bias detection and more rapid remediation when problems emerge.

The challenge for organizational leaders lies in balancing competing pressures, innovation speed versus careful deliberation, competitive advantage versus transparency, and efficiency gains versus employment impacts. Amazon's decision to discontinue its AI recruiting tool after discovering gender bias illustrates leadership choosing ethical integrity over operational convenience, even when the decision carried costs. This case demonstrates that ethical leadership sometimes requires abandoning investments and admitting mistakes rather than deploying flawed systems.

Building organizational capacity for ethical AI requires sustained investment in education and culture change. Leaders must ensure that teams possess not only technical skills but also ethical literacy, the ability to recognize when AI applications raise concerns about fairness, privacy, autonomy, or dignity. Organizations that integrate ethics training into professional development, reward employees who raise concerns, and protect those who challenge problematic projects create environments where ethical considerations become routine rather than exceptional. The future of ethical AI depends substantially on whether organizational leaders choose to build these capabilities now, while AI systems are still being shaped, or defer action until patterns become entrenched and harder to change.

Advocacy and Policy Engagement: Making Your Voice Heard in AI Governance

The development of artificial intelligence governance frameworks is not happening in distant boardrooms beyond public reach; it is occurring in city councils, state legislatures, regulatory agencies, and international bodies where citizen voices can and do influence outcomes. Research from the Brookings Institution demonstrates that AI policy is increasingly shaped by those most affected by its outcomes, not solely by technologists or policymakers in isolation. The question is whether individuals will choose to participate in these governance processes or allow decisions to be made without their input.

Policy engagement in AI governance takes multiple forms, each offering distinct pathways for influence. Public comment periods on proposed regulations provide formal mechanisms for individuals and organizations to submit feedback that agencies are legally required to consider. When the Federal Trade Commission sought input on AI and algorithmic transparency in 2020, thousands of comments from citizens, advocacy groups, and affected communities shaped the agency's subsequent enforcement priorities. These comments documented real-world harms, challenged industry claims, and provided evidence that influenced regulatory thinking in measurable ways.

Legislative advocacy represents another critical avenue for shaping AI governance. Contacting elected representatives, testifying at hearings, and supporting legislation that protects human rights in AI contexts directly influences which bills advance and which provisions they contain. The successful campaigns to ban government use of facial recognition technology in San Francisco, Boston, and other municipalities emerged from sustained civic engagement by ordinary citizens who attended public meetings, organized community opposition, and demanded accountability from elected officials. Research from **MIT's** Media Lab shows that public resistance to biased or invasive AI applications has directly influenced corporate decisions to delay or modify product launches, proving that citizen voices aggregated through civic channels can alter technological development.

Building coalitions amplifies individual voices and creates sustained pressure for policy change. The **AI Ethics Council at Operation HOPE** exemplifies how multidisciplinary coalitions bring together experts from technology, business, government, and civil society to guide ethical principles for AI and promote inclusive policy development. These coalitions provide frameworks for collective action, coordinate advocacy efforts across organizations, and ensure that diverse perspectives, particularly from marginalized communities, are represented in policy discussions. **UNESCO's** Recommendation on the Ethics of Artificial Intelligence, adopted by over **190 countries**, emerged from extensive stakeholder engagement that included civil society organizations, academic institutions, and community representatives alongside government officials.

Effective policy advocacy requires both technical literacy and strategic communication. Understanding how AI systems work, what data they collect, and what rights individuals possess enables more persuasive advocacy grounded in evidence rather than abstract concerns. Organizations like Globethics and PoliSync have developed practical resources for nonprofits and community groups to engage with AI policy, emphasizing that advocacy needs not require advanced technical expertise but does demand clear articulation of values and impacts. Studies published in Nature Machine Intelligence indicate that even basic AI literacy significantly increases individuals' capacity to participate meaningfully in governance discussions and identify problematic applications requiring policy intervention.

The urgency of policy engagement stems from the reality that governance frameworks being established now will shape AI's trajectory for decades. **Dr. Nicol Turner Lee** of the Brookings Institution emphasizes that advocacy is not just about influencing policy; it is about ensuring that the voices of marginalized communities are heard in the design and governance of AI systems. Once regulatory precedents are set, industry practices become entrenched, and technical standards are adopted, changing courses becomes exponentially more difficult. The window for shaping foundational AI governance remains open, but it requires citizens who understand that their participation matters and who are willing to invest time and effort in making their voices heard through formal channels, coalition building, and sustained engagement with policymakers at every level of government.

Education and Digital Literacy: Preparing Ourselves and Future Generations

The most powerful tool for shaping AI's ethical trajectory may be the least technological: education. As artificial intelligence systems increasingly permeate healthcare, employment, criminal justice, and daily life, the capacity of individuals to understand, question, and influence these technologies becomes essential to preserving human dignity and democratic participation. 4 Research from MIT's Media Lab demonstrates that societies with higher levels of digital literacy show greater resistance to algorithmic manipulation, more effective advocacy for privacy protections, and stronger demand for transparency in AI systems. The question is not whether education matters in the AI age, but whether we will invest in building the literacy required for meaningful engagement before patterns of technological deployment become irreversible.

Digital literacy in the context of AI extends far beyond basic computer skills or the ability to use applications. It encompasses understanding how machine learning systems make decisions, recognizing when algorithms may embed bias, grasping the implications of data collection and surveillance 67, and possessing the conceptual frameworks needed to evaluate ethical trade-offs. The Stanford Institute for Human-Centered Artificial Intelligence emphasizes that effective AI literacy requires interdisciplinary knowledge drawing from computer science, ethics, law, and social sciences 67, a synthesis that traditional educational systems have been slow to provide.

The current educational landscape reveals troubling gaps. Most K-12 curricula lack comprehensive coverage of AI fundamentals or ethical considerations 67, leaving students unprepared for a world where algorithmic systems will shape their opportunities, privacy, and autonomy. A 2022 study published in Computers & Education found that fewer than fifteen percent of U.S. high schools offered any coursework addressing AI ethics or algorithmic bias, even though students already encounter these systems daily through social media, educational platforms, and college admissions processes. This gap creates generations of citizens who consume AI-powered services without the knowledge needed to recognize manipulation, demand accountability, or participate meaningfully in governance decisions.

Higher education faces similar challenges. While computer science programs increasingly include ethics modules, students in other disciplines, from medicine and law to business and social work, often graduate without exposure to AI's implications for their fields. Healthcare professionals make treatment decisions influenced by algorithmic recommendations without training in how to evaluate these systems' reliability or bias. 4 Lawyers encounter AI-generated legal research without understanding its limitations. Business leaders implement AI systems affecting thousands of employees without frameworks for assessing ethical impacts. 2 The IEEE Global Initiative on Ethics of Autonomous and Intelligent Systems argues that AI literacy must become a core competency across professions, not an optional specialization for technologists alone.

Effective AI education requires age-appropriate progression that builds critical thinking alongside technical understanding. Elementary students can learn basic concepts of fairness and how computers make decisions. Middle school curricula can introduce algorithmic bias, data privacy, and the distinction between correlation and causation. High school students can engage with case studies examining real-world AI failures and successes 3, 123, developing the analytical skills needed to evaluate competing claims about technology's benefits and risks. Research from the University of Helsinki's AI Education program demonstrates that even young children can grasp fundamental concepts when presented through concrete examples and hands-on activities, building intuition that serves as a foundation for more sophisticated understanding.

Professional education and lifelong learning represent equally critical frontiers. Given AI's rapid evolution, one-time training becomes obsolete quickly. Organizations must invest in ongoing education for employees who develop, deploy, or are affected by AI systems. **2** The European Union's Digital Education Plan emphasizes continuous learning as essential for maintaining democratic participation in an AI-saturated society, recommending that member states establish accessible programs for citizens at all career stages.

Building widespread digital literacy faces significant obstacles: limited educational resources, rapidly changing technology, and unequal access across socioeconomic and geographic boundaries. Yet these challenges make the effort more urgent, not less. The alternative to investing in education is a future where AI systems are understood and controlled by a narrow technical elite while the majority of humanity experiences their effects without the knowledge needed to shape their development or demand accountability. **4**

Building Coalitions and Communities: Collective Action for Ethical AI

The transformation of artificial intelligence from a narrow technical concern into a force reshaping society demands responses that match its scale and complexity. Individual choices matter, organizational leadership creates change, and policy engagement influences governance, but none of these pathways alone possess sufficient power to ensure AI serves humanity's collective interests. **124, 126** The challenges posed by AI ethics require coordinated action among diverse stakeholders who recognize that their shared future depends on collaboration rather than competition or isolation. **126**

Collective action in AI ethics addresses a fundamental problem: when individual interests diverge from collective interests, the result often favors those with the most resources and power while distributing harm broadly across society. 124 demonstrates that many AI-related decisions present classic collective action problems, where competitive pressures incentivize rushing development to achieve first-mover advantage, potentially compromising safety and ethical considerations. 126 Without coordinated responses, the benefits of AI remain concentrated among privileged actors while vulnerable communities experience marginalization and harm. 124

The Collective Impact Coalition for Ethical Artificial Intelligence exemplifies how coordinated stakeholder engagement can produce measurable progress. Launched in 2022, this coalition comprises investors and civil society organizations working together to encourage technology companies to advance ethical AI policies and practices. 125 The coalition operates with three primary objectives: raising awareness about responsible AI, increasing understanding of best practices, and improving corporate commitments to ethical principles. 125 Within the first year following the campaign's launch, nineteen companies evaluated in the Digital Inclusion Benchmark published their inaugural ethical AI principles, demonstrating that coordinated pressure from multiple stakeholders can shift corporate behavior in ways that individual voices cannot achieve alone. 125

Effective coalitions extend beyond formal investor campaigns to include community-centered research organizations and advocacy groups that bridge knowledge gaps and amplify marginalized voices. The Distributed AI Research Institute conducts community-centered research that reflects lived experiences to inform the development, deployment, and access to AI assets. 124 The Algorithmic Justice League advocates for equitable and accountable AI using art and research to illuminate discrimination potential, engage with policymakers and industry to prevent harm, and amplify voices of those most affected by algorithmic systems. 124 These organizations demonstrate that collective action succeeds when it incorporates diverse perspectives, including those of developers, ethicists, artists, advocates, and community representatives, thereby ensuring that technical expertise combines with lived experience and ethical reasoning. 124

Building effective coalitions requires strategic investment in three interconnected areas. First, engaging creative professionals, designers, musicians, filmmakers, and photographers in collaboration with diverse voices and visionary funders helps design AI that reflects varied cultural perspectives and addresses local social issues. 124 Second, supporting intermediary organizations that serve as connective tissue among funders, organizations, and communities facilitates collaborative knowledge exchange, sharing of best practices, and ongoing dialogue about ethical challenges. 124 Third, reimagining AI as public infrastructure rather than proprietary technology shifts focus from private interests to collective good, prioritizing accessibility and ensuring benefits are shared across diverse communities. 124

Academic research on collective action problems identifies three primary solution types: government regulation, private markets, and community self-organization. 126 Each approach offers distinct strengths, and no single approach constitutes a complete solution. 126 Government regulation establishes binding standards, private markets incentivize ethical practices through consumer preference and investor pressure, and community self-organization fosters grassroots innovation and local accountability. 126 A combination of these approaches, operating at different scales and addressing different aspects of AI governance, provides the most robust framework for ensuring ethical development. 126

The responsibility of civil society in AI governance extends beyond defending against harms to actively reimagining how communities live together and care for each other, then implementing ethical AI applications to realize these visions. 124 This requires long-term, unrestricted investments in cooperatives, communities, and collaborative networks that transcend borders, sectors, and issues. 124 The institutions developed now to resolve AI collective action problems will have lasting consequences for how AI shapes human futures, making the choice to build collaboratively one of the most consequential decisions facing contemporary society. 126

Living with Intention in an AI World: Daily Practices for Ethical Technology Use

The transformation of AI from abstract technology into daily reality requires more than awareness; it demands intentional practice. While advocacy, education, and organizational change shape AI's broader trajectory, the choices individuals make each day about how they engage with technology determine whether AI enhances human flourishing or diminishes it. 4, 67 Research from Stanford's Institute for Human-Centered Artificial Intelligence demonstrates that deliberate habits around technology use significantly influence both individual well-being and the collective signals that shape corporate behavior and product development.

Living with intention in an AI-saturated world begins with mindful engagement rather than passive consumption. Before using AI-powered tools, recommendation engines, voice assistants, and automated decision systems, individuals can pause to consider their purpose and potential impacts. This practice of critical consumption involves questioning how recommendations are generated, whose interests they serve, and what data they rely upon. 4, 67 Studies published in Computers in Human Behavior indicate that individuals who regularly question AI outputs demonstrate greater resistance to algorithmic manipulation and make more autonomous decisions aligned with their values rather than system-optimized outcomes.

Data privacy and consent awareness constitute another essential daily practice. The **2022** phenomenon documented by researchers as the "privacy paradox" reveals that individuals express concern for digital privacy yet routinely disclose personal information to AI-powered platforms for convenience. **67** However, interventions as simple as daily reminders to review privacy settings or reflect on data usage significantly improved digital habits and reduced risky sharing behaviors. **67** Taking time to understand how personal data is collected, stored, and used by AI systems **2, 67**, and using privacy settings to limit unnecessary data collection **67**, represents a concrete practice that protects individual autonomy while sending market signals about consumer expectations for data stewardship.

Cultivating digital literacy and skepticism serves as ongoing protection against AI-enabled harms. Staying informed about AI capabilities and limitations, recognizing common issues such as algorithmic bias and data misuse **2, 3**, and maintaining vigilance about AI-generated content and deepfakes **6, 7** constitute daily practices that preserve critical thinking in an environment designed to optimize engagement over accuracy. The MIT Media Lab emphasizes that even basic AI literacy significantly increases individuals' capacity to detect manipulation, evaluate competing claims, and make informed decisions about which technologies to trust.

Practicing fairness and inclusion through daily technology choices creates pressure for more equitable systems. When encountering AI systems, individuals can consider who may be disadvantaged by their design or outputs, provide feedback to developers about unfair or discriminatory effects **67, 68,** and favor platforms that demonstrate commitment to ethical design and inclusivity **4, 67**. The collective impact of these individual choices becomes visible in market dynamics; when consumers systematically choose ethical alternatives, companies respond. Research from the Berkman Klein Center at Harvard demonstrates that consumer preferences for privacy-protective and bias-mitigated systems have influenced product development priorities at major technology companies.

Demanding transparency and accountability in daily interactions with AI systems reinforces the principle that automated decisions affecting human lives require explanation and oversight. Requesting explanations when AI systems make consequential decisions, holding providers accountable for opaque practices **2, 67**, and advocating for ethical standards in public dialogues about responsible AI **67** represent practices that individuals can integrate into routine technology use. The ACM Code of Ethics emphasizes that users bear responsibility not only for their own choices but also for creating environments that make ethical technology development economically and socially advantageous.

These daily practices do not require technical expertise or extraordinary effort; they require consciousness, consistency, and the recognition that technology's future is not predetermined but shaped by the accumulated choices of millions of individuals who refuse to accept that their habits do not matter. The world we want emerges not from grand gestures alone but from the intentional decisions made each day about how we engage with the systems increasingly mediating human experience. The trajectory of artificial intelligence is not predetermined by technological inevitability or corporate ambition. It is being written through the accumulated choices of individuals, organizations, and communities who recognize that their actions matter. Throughout this chapter, the pathways for influence have been mapped across multiple domains: personal technology choices that send market signals, organizational leadership that embeds ethics into development processes, policy engagement that shapes governance frameworks, educational investments that build critical literacy, coalition building that amplifies collective power, and daily practices that preserve human autonomy in an AI-saturated world. Each pathway alone possesses limited power, but together they constitute a comprehensive strategy for ensuring that artificial intelligence serves humanity rather than diminishes it.

The urgency of action stems from a fundamental reality about technological systems: they become progressively harder to change as they mature. The AI systems being deployed today establish precedents, generate training data for future models, create economic dependencies, and build infrastructure that becomes entrenched over time. Research from Stanford's Institute for Human-Centered Artificial Intelligence demonstrates that interventions addressing bias, privacy, and accountability prove far more effective when implemented during initial development than when retrofitted into established systems. The window for shaping AI's foundational architecture remains open, but it will not remain open indefinitely. The choices made in the next decade will determine whether AI amplifies human dignity or erodes it, whether benefits are shared equitably or concentrated among the privileged, and whether democratic societies retain meaningful control over technologies that increasingly mediate human experience.

The question confronting every reader is not whether they possess the expertise to influence AI's future; it is whether they will choose to exercise the agency they already hold. Individuals need not be data scientists to demand transparency, organizational leaders need not be ethicists to prioritize human welfare over efficiency, and citizens need not be policymakers to participate in governance processes. What is required is the recognition that artificial intelligence reflects human values and that those values will be determined by who shows up to shape them. History demonstrates repeatedly that technological trajectories shift when diverse voices insist on alternatives to the paths of least resistance, when professionals refuse to implement systems that violate their principles, and when communities organize around shared visions of human flourishing.

The world we want, where AI enhances human capability without eroding human dignity, where innovation proceeds with wisdom rather than reckless speed, and where technology's benefits reach those most in need rather than only those already privileged, will not emerge from passiveness or hope alone. It requires the conscious, sustained effort of people who understand that the future is not something that happens to us but something we create through deliberate choice. The frameworks, strategies, and practices outlined in this chapter provide concrete starting points, but their effectiveness depends entirely on whether readers transform knowledge into action.

Standing at this pivotal moment, humanity faces a choice that will echo for generations. We can accept whatever AI future emerges from the priorities of those currently holding power, or we can actively participate in building systems that reflect our highest values and deepest commitments to human dignity, equity, and flourishing. The difference between these futures comes down to the actions people take in critical moments, and this is such a moment. The garden of our technological future, like any garden, will grow what we choose to plant and tend. The question is whether we will pick up the tools and begin the work.

CONCLUSION

The future doesn't simply happen to us. The difference between tragedy and recovery often comes down to the actions people take in critical moments, and we are living through such a moment now as artificial intelligence reshapes the foundations of human society.

The central truth that has threaded through every chapter of this book remains unchanged: AI itself carries no moral weight. The algorithms that diagnose diseases, recommend treatments, approve loans, screen job candidates, and increasingly shape our daily experiences are neither good nor evil; they are reflections of human choices, human values, and human responsibility. When an AI system discriminates, it does so because humans created biased training data or failed to test for fairness. When algorithms invade privacy, it happens because humans prioritize profit or convenience over dignity. When AI threatens truth or amplifies inequality, these outcomes stem from human decisions about what to build, how to deploy it, and whom to protect.

This recognition should be empowering rather than paralyzing. We have not yet crossed into a future where AI operates beyond human control or understanding. The systems being deployed today, the regulations being written now, and the norms being established in this decade will determine whether artificial intelligence becomes a force that elevates human dignity or diminishes it. The world we want is still within reach, but only if we act with intention and urgency.

The ethical challenges explored throughout this book, bias and fairness, privacy and surveillance, accountability and governance, and equitable access to AI-enhanced medicine, are not abstract philosophical puzzles.

They manifest in real life every day. A person was denied a job interview because an algorithm screened out their resume. A patient is receiving substandard care because a healthcare algorithm underestimated their needs. A community subjected to invasive surveillance that treats entire populations as suspects. These harms are preventable, but only through conscious effort.

The practical frameworks and insights offered in these chapters provide pathways forward. Responsible AI development that embeds ethics from conception to deployment. Privacy-preserving technologies that protect human dignity while enabling innovation. Governance structures that establish clear accountability and give affected communities a voice. Medical AI applications that expand access rather than deepening inequality. These solutions exist, but they require commitment from individuals, organizations, and societies willing to prioritize human welfare over efficiency or profit.

Hope is not passive optimism; it is active engagement with reality as it is, combined with determined effort to shape what comes next. The AI systems analyzing my brain scans, potentially accelerating drug discovery, and offering new therapeutic possibilities represent genuine hope for millions facing chronic conditions. But that hope depends entirely on whether we ensure these technologies serve all patients with equity and compassion, not just those with resources and access.

Each reader of this book holds influence in multiple spheres. As professionals implementing AI systems, as citizens engaging with policy, as consumers making technological choices, as community members advocating for ethical practices, and as human beings modeling values for the next generation. The question is not whether AI will shape our future; it already is. The question is whether we will shape AI with the ethical clarity, moral courage, and collective wisdom that this moment demands.

We still have time to choose the world we want. But time is not infinite, and the choices we make now will echo across generations. The garden we tend today determines what grows tomorrow. Let us tend it with care, with ethics, with unwavering commitment to human dignity, and with the understanding that this responsibility belongs to all of us.

If you enjoyed this book, I would be incredibly grateful if you could take a moment to leave an honest review on Amazon. Your feedback not only helps other readers discover the book, but it also supports independent authors like me in continuing to create meaningful work.

YOUR VOICE TRULY MATTERS

If the book helped you learn something new, feel more empowered about your health, or simply sparked your curiosity, sharing that experience makes a meaningful difference. Simply scan the QR code.

Thank you so much for your time, your support, and for being part of this artificial intelligence journey.

With Gratitude,
Alexandria Isaacs.

BIBLIOGRAPHY

- Santos, O. (2024, February 09). Beyond the Algorithm: AI, Security, Privacy, and Ethics. Barnes & Noble.
- Santos, O. (2024, January 30). Beyond the Algorithm: AI, Security, Privacy, and Ethics. Goodreads.
- Sterling, A. (2025, April 01). Beyond the Algorithm: Understanding AI Ethics. Barnes & Noble.
- Lavanchy, M. (2018, November). Amazon's sexist hiring algorithm could still be better than a human. IMD.
- Goodman, R. (2018, October 12). Why Amazon's Automated Hiring Tool Discriminated Against Women. ACLU.
- Parsani, P. (2024, June 6). Case Study: How Amazon's AI Recruiting Tool "Learnt" Gender Bias. Cut the SaaS.
- The Regulatory Review. (2021, September 11). Saturday Seminar: Countering Bias in Algorithmic Hiring Tools. The Regulatory Review.
- Smith Brain Trust. (2018, October 31). The Problem with Amazon's AI Recruiter. University of Maryland Robert H. Smith School of Business.
- Smith, D. (2024, October 30). Accounting for bias in medical data helps prevent AI from amplifying racial disparity. Michigan Engineering News.
- Meline, M. (2024, April 25). Health Care Algorithms Can Improve or Worsen Racial and Ethnic Disparities. Penn LDI.
- Stetler, C. (2024, November 14). AI Algorithms Used in Healthcare Can Perpetuate Bias. Rutgers University-Newark.
- Bosco, C. M., Chin, M. H., & Parker, W. F. (2025, Spring). Addressing Bias in Clinical Algorithms to Advance Health Equity. AFT Health Care.
- Nigam, A. (2024, November 13). How I addressed racial bias in my company's AI algorithm. STAT.
- Stonier, J., Woodman, L., & Yee Amezaga, K. (2024, September). The lessons we're learning from addressing racial bias in healthcare algorithms can and should inform how we approach data equity across other sectors. World Economic Forum.
- Cedars-Sinai. (2025, June 30). Cedars-Sinai Study Shows Racial Bias in AI-Generated Treatment Regimens for Psychiatric Patients. Cedars-Sinai Newsroom.

- Joseph, J. (2025, July 23). Algorithmic bias in public health AI: a silent threat to equity in low-resource settings. Frontiers in Public Health.
- Sedlarcik, S. (2024, November 05). Ethical AI Sentencing: A Framework for Moral Judgment in Criminal Justice. Critical Debates HSGJ.
- Council on Criminal Justice. (2025, April). DOJ Report on AI in Criminal Justice: Key Takeaways. Council on Criminal Justice.
- Bernstein, G. (2024, September 27). The Ethics of AI Decision-Making in the Criminal Justice System. CloudTweaks.
- Bigelow, S. J., Gillis, A. S., & Pratt, M. K. (2024, September 09). What is machine learning bias (AI bias)? TechTarget.
- Greene, J. (2024). Algorithmic bias. EBSCO Research Starters.
- Chen, J., & Burch, C. (2025, October 10). Bias & Fairness in AI Models. Contrary Research.
- Simons, T. (2020, November 30). Addressing issues of fairness and bias in AI. Thomson Reuters Institute.
- Demirchyan, G. (2025, August 29). Algorithmic fairness: challenges to building an effective regulatory regime. Frontiers in Artificial Intelligence.
- Ladd, J. (2022, April). Discrimination By Proxy: How AI Uses Big Data to Discriminate. MTLR.
- Prince, A. E. R., & Schwarcz, D. (2020, March 15). Proxy Discrimination in the Age of Artificial Intelligence and Big Data. Iowa Law Review.
- Bauer, K., & Gill, A. (2023, May 3). Mirror, Mirror on the Wall: Algorithmic Assessments, Transparency, and Self-Fulfilling Prophecies. Information Systems Research.
- 39 Mishor, A. (2022, January). Stop Creating Self-Fulfilling Prophecies: How to Apply AI to Small Data Problems. Leadspace.
- Mertens, M. (2025, June 5). Genetic tests as prophecy: understanding self-defeating and self-fulfilling mechanisms in (predictive) genetic testing. PubMed Central.
- King, O. C., & Mertens, M. (2023). Self-fulfilling Prophecy in Practical and Automated Prediction. PhilArchive.
- ANDOPEN. (2025). The State of Facial Recognition Technology in 2025: Accuracy, Performance, and Future Trends. AN OPEN.
- 45 CreateBytes. (2025, November 18). Facial Recognition in 2025: The Complete Guide to Technology, Trends & Ethics. CreateBytes.
- Security Industry Association. (2024, March 11). What Science Really Says About Facial Recognition Accuracy and Bias Concerns. Security Industry Association.
- Rewari, V. (2024, September 28). AI Bias Audit: 7 Steps to Detect Algorithmic Bias. OptiBlack.

- Bronsdon, C. (2025, April 7). Detecting and Mitigating Model Biases in AI Systems. Galileo.
- Gray, M., Samala, R., Liu, Q., Skiles, D., Xu, J., Tong, W., & Wu, L. (2023, November 29). Measurement and Mitigation of Bias in Artificial Intelligence: A Narrative Literature Review for Regulatory Science. Clinical Pharmacology & Therapeutics.
- Zdrok, O. (2024, June 4). Fairness Metrics in AI: Your Step-by-Step Guide to Equitable Systems. Shelf.
- Laenens, W. (2025, March 25). (EN) What can ethics-by-design mean for responsible AI innovation? 5 key takeaways. Knowledge Centre Data & Society.
- Sustainability Directory. (2025, October 11). Algorithmic Impact Assessments. Sustainability Directory.
- Ada Lovelace Institute. (2022, February 8). Algorithmic impact assessment: user guide. Ada Lovelace Institute.
- Stahl, B. C. (2023, March 24). A systematic review of artificial intelligence impact assessments. PubMed Central.
- Government of Canada. (2025, October 10). Algorithmic Impact Assessment tool. Canada.ca.
- Data & Society. (2023, May). Algorithmic Impact Methods Lab. Data & Society.
- Watkins, E. A., Moss, E., Metcalf, J., Singh, R., & Elish, M. C. (2021, July). Governing Algorithmic Systems with Impact Assessments: Six Observations. ACM Digital Library.
- AI Guru. (2025, February 11). AI Ethics: Artificial Intelligence Beyond the Algorithm. Google Books.
- Vats, N. (2024, January 1). The Ethics of AI in Monitoring and Surveillance. NICE Actimize.
- Radanliev, P. (2025, June 17). Privacy, ethics, transparency, and accountability in AI systems for wearable devices. Frontiers in Digital Health.
- Office of the Director of National Intelligence. (2020, June). Artificial Intelligence Ethics Framework for the Intelligence Community. Intelligence.gov.
- Lynch, K. (2025, August 21). Ethical AI Principles: Fairness, Transparency, and Trust in HR. Phenom.
- LunarTech. (2025, January 18). AI-Driven Predictive Analytics for Healthcare. LunarTech.
- Williamson, D. (May 28, 2025). Where AI Meets Health Data Privacy: How to Balance Innovation with Risk. Datavant.
- Pihtovnicov, A. (2025, June 03). AI Predictive Analytics in Healthcare: What Is It & Benefits. TechMagic.
- Masta, L. (2024, November 30). A complete overview of predictive analytics in healthcare. Arcadia.

- Dixon, D. (2024, May 9). Unveiling the Influence of AI Predictive Analytics on Patient Outcomes: A Comprehensive Narrative Review. Curious.
- Park University. (2024, December 2). AI in Healthcare: Enhancing Patient Care and Diagnosis. Park University.
- Sun, H. (2025, March 19). Bridging the Accountability Gap in AI Decision-Making: An Integrated Analysis of Legal Precedents and Scholarly Perspectives. Frontiers in Humanities and Social Sciences.
- Thomas, M. (2025). The AI Accountability Gap: Why Boards Must Invest in Verification Capability. Mario Thomas.
- Stanford, E. (2025, May 15). Autonomous AI: who is responsible when AI acts autonomously and does things go wrong? Global Legal Insights.
- Mack, O. V. (2024, March). Legal Tech: Liability and Responsibility in the Age of AI. ACC Docket.
- Golan, G. (2025, May 20). AI Governance Framework: Key Principles & Best Practices. Data Privacy Hub.
- Short, L. (2025, June 26). Building a Responsible AI Framework: 5 Key Principles for Organizations. Harvard Division of Continuing Education
- Young, F. (2025, October 13). AI as a Product: The Next Frontier in Product Liability Law. UIC Law Library.
- National Telecommunications and Information Administration. (2024, March 27). Liability Rules and Standards. National Telecommunications and Information Administration.
- Howarth, M. (2025, January 7). AI liability—who is accountable when artificial intelligence malfunctions? Taylor Wessing.
- Frazier, K. (2025, June 4). We're Not Ready for AI Liability. AI Frontiers.
- Valley, T. S., Kazerooni, E. A., Banovic, N., Wiens, J., & Sjoding, M. W. (2023). Measuring the Impact of AI in the Diagnosis of Hospitalized Patients: A Randomized Clinical Vignette Survey Study. JAMA.
- Akinrinmade, A. O. (2023, September 20). Artificial Intelligence in Healthcare: Perception and Reality. Curious.
- Ogbonna, N. (2025, April). How AI Achieves 94% Accuracy in Early Disease Detection: New Research Findings. GlobalRPh.
- RamSoft. (2025, May 16). Understanding the Accuracy of AI in Diagnostic Imaging. RamSoft.
- Johnson, K. B., Wei, W., Weeraratne, D., Frisse, M. E., Misulis, K., Rhee, K., Zhao, J., & Snowdon, J. L. (2020, October 12). Precision Medicine, AI, and the Future of Personalized Health Care. PMC - PubMed Central.
- Bajwa, J., Munir, U., Nori, A., & Williams, B. (2021, July). Artificial intelligence in healthcare: transforming the practice of medicine. Future Healthcare Journal.

- Mashar, M., Chawla, S., Chen, F., Lubwama, B., Patel, K., Kelshiker, M. A., Bachtiger, P., & Peters, N. S. (2023, January 16). Artificial Intelligence Algorithms in Health Care: Is the Current Food and Drug Administration Regulation Sufficient? JMIR AI.
- Thoughtful AI. (2025, August 26). AI in Medicine: Transforming Patient Treatment and Care. Thoughtful AI.
- Johns Hopkins Bloomberg School of Public Health. (2023). How Health Care Algorithms and AI Can Help and Harm. Johns Hopkins Bloomberg School of Public Health.
- Angus, D. C. (2025). AI, Health, and Health Care Today and Tomorrow: The JAMA Summit Report on Artificial Intelligence. JAMA Network.
- Scripps Research. (2025, May 29). AI pinpoints new anti-aging drug candidates. Scripps Research.
- Oreskovic, A. (2025, October 30). Longevity science is on the cusp of major breakthroughs thanks to AI, but significant 'data gaps' need to be filled, an expert says. Fortune.
- Insilico Medicine. (2025, August 21). ARDD for healthy longevity: How science, technology, and investment are redefining the future. In silico Medicine.
- Wilczok, D. (2025, January 16). Deep learning and generative artificial intelligence in aging research and healthy longevity medicine. Aging.
- Santos, O., & Radanliev, P. (2024, June 17). Beyond the Algorithm: AI, Security, Privacy, and Ethics. Audible.
- Sharko, M., & Cole, C. L. (2025, March 11). Integrating Artificial Intelligence Support in Patient Care While Respecting Ethical Principles. JAMA Network Open.
- World Health Organization. (2023, May 16). WHO calls for safe and ethical AI for health. World Health Organization.
- UCSF. (2025, February). How Artificial Intelligence Is Fostering Human Connection in Health Care. UCSF.
- Farhud, D. D. (2021, November). Ethical Issues of Artificial Intelligence in Medicine and Healthcare. Iranian Journal of Public Health.
- Sisk, B. A., Before, A., Burgers, J., & DuBois, J. M. (2020). How Will Artificial Intelligence Affect Patient-Clinician Relationships? . AMA Journal of Ethics.
- Yu, S., Lee, S., & Hwang, H. (2024, December 6). The ethics of using artificial intelligence in medical research. Kosin Medical Journal.
- Santos, O. & Radanliev, P. (2024, February 9). Beyond the Algorithm: AI, Security, Privacy, and Ethics. Murder By the Book.
- Srivastava, L. (2024). Building Community Governance for AI. Stanford Social Innovation Review.

REFERENCES

Bostrom, Nick. *Superintelligence: Paths, Dangers, Strategies.* Oxford University Press, 2014.

Bryson, Joanna J. "The Artificial Intelligence Ethics Divide." *Communications of the ACM*, vol. 61, no. 7, 2018, pp. 32–39.

Crawford, Kate. *Atlas of AI: Power, Politics, and the Planetary Costs of Artificial Intelligence.* Yale University Press, 2021.

Floridi, Luciano. *The Ethics of Information.* Oxford University Press, 2013.

Goodfellow, Ian, Yoshua Bengio, and Aaron Courville. *Deep Learning.* MIT Press, 2016.

Marcus, Gary, and Ernest Davis. *Rebooting AI: Building Artificial Intelligence We Can Trust.* Pantheon, 2019.

O'Neil, Cathy. *Weapons of Math Destruction.* Crown, 2016.

Russell, Stuart. *Human Compatible: Artificial Intelligence and the Problem of Control.* Viking, 2019.

Tegmark, Max. *Life 3.0: Being Human in the Age of Artificial Intelligence.* Alfred A. Knopf, 2017.

Made in the USA
Monee, IL
04 November 2026

67384800R00095